Maria Loredana Kehrer

Thermomechanical Mean-Field Modeling and Experimental Characterization of Long Fiber-Reinforced Sheet Molding Compound Composites

Schriftenreihe
Kontinuumsmechanik im Maschinenbau
Band 15

Karlsruher Institut für Technologie (KIT)
Institut für Technische Mechanik
Bereich Kontinuumsmechanik

Hrsg. Prof. Dr.-Ing. habil. Thomas Böhlke

Eine Übersicht aller bisher in dieser Schriftenreihe erschienenen Bände finden Sie am Ende des Buchs.

Thermomechanical Mean-Field Modeling and Experimental Characterization of Long Fiber-Reinforced Sheet Molding Compound Composites

by
Maria Loredana Kehrer

KIT Scientific Publishing

Dissertation, Karlsruher Institut für Technologie
KIT-Fakultät für Maschinenbau

Tag der mündlichen Prüfung: 14. Dezember 2018
Referenten: Prof. Dr.-Ing. Thomas Böhlke, Prof. Dr. Jeffrey T. Wood

Impressum

KIT Scientific Publishing

Karlsruher Institut für Technologie (KIT)
KIT Scientific Publishing
Straße am Forum 2
D-76131 Karlsruhe

KIT Scientific Publishing is a registered trademark
of Karlsruhe Institute of Technology.
Reprint using the book cover is not allowed.

www.ksp.kit.edu

Print on Demand 2019 – Gedruckt auf FSC-zertifiziertem Papier

ISSN 2192-693X
ISBN 978-3-7315-0924-0
DOI 10.5445/KSP/1000093328

Thermomechanical Mean-Field Modeling and Experimental Characterization of Long Fiber-Reinforced Sheet Molding Compound Composites

Zur Erlangung des akademischen Grades

Doktor der Ingenieurwissenschaften

der KIT-Fakultät für Maschinenbau

Karlsruher Institut für Technologie (KIT)

genehmigte

Dissertation

von

M.Sc. Maria Loredana Kehrer

Tag der mündlichen Prüfung:	14.12.2018
Hauptreferent:	Prof. Dr.-Ing. Thomas Böhlke
Korreferent:	Prof. Dr. Jeffrey T. Wood

Zusammenfassung

Faserverstärkte Polymere tragen wesentlich zu gewichtsreduzierten Bauteilkonstruktionen im Automobil- und Luftfahrtsektor bei. Aufgrund der Fließfähigkeit diskontinuierlich faserverstärkter Polymere (DiCoFRP) während des Herstellungsprozesses können komplexe Bauteilgeometrien erzielt werden. Im Gegensatz zu DiCoFRP erreichen kontinuierlich faserverstärkte Polymere (CoFRP) durch ihren hohen Faservolumengehalt eine hohe spezifische Steifigkeit und Festigkeit. Das Ziel der Forschungsarbeiten des internationalen Graduiertenkollegs DFG-GRK 2078 ist es, die Vorteile von DiCoFRP und CoFRP zu kombinieren. Die neu entwickelte Materialklasse zeichnet sich durch die Integration von Funktionselementen in den Herstellungsprozess, eine hohe Designfreiheit von Bauteilen und eine wirtschaftliche Großserienfertigung aus. Das übergeordnete Ziel des deutsch-kanadischen Konsortiums dieses Graduiertenkollegs ist es, Konzepte zur Charakterisierung, Modellierung, Dimensionierung und Verarbeitung von DiCoFRP und CoFRP sowie der Kombination beider Materialsysteme zu entwickeln.

In der vorliegenden Arbeit wird ein diskontinuierlich faserverstärktes Duroplast untersucht, das mit dem Sheet Molding Compound (SMC) Verfahren hergestellt wurde. Aufgrund der prozessbedingten Faserorientierungsverteilung entsteht ein Komposit mit anisotroper Mikrostruktur, die die mechanischen Eigenschaften bestimmt.

Die zentralen Zielstellungen dieser Arbeit sind die Modellierung des thermoelastischen Materialverhaltens des Komposites unter Berücksichtigung der vorliegenden Mikrostruktur sowie die experimentelle Charakterisierung des Reinharzes und des Komposites.

Zur experimentellen Charakterisierung der thermoviskoelastischen Materialeigenschaften, wird die dynamisch mechanische Analyse (DMA) verwendet. Basierend auf diesen Tests werden materialspezifische Eigenschaften, wie der temperaturabhängige E-Modul oder die Glasübergangstemperatur ermittelt. Zudem werden Messungen zur Bestimmung des thermischen Ausdehnungskoeffizienten des reinen Matrixmaterials und des Verbundwerkstoffs durchgeführt. Die experimentellen Daten des Matrixmaterials dienen als Eingangsdaten für die Materialmodelle. Die numerisch ermittelten Materialparameter werden mit den experimentellen Daten des Verbundwerkstoffs verglichen.

Zur Approximation des effektiven thermoelastischen Materialverhaltens wird die Hashin-Shtrikman Mean-Field Homogenisierungsmethode, basierend auf Willis (1981) und Walpole (1966), in der Formulierung mit Eigendehnungen verwendet. Die effektiven Größen werden in zwei Schritten bestimmt. Hierbei ist die Homogenisierungsmethode in Abhängigkeit einer variablen Referenzsteifigkeit formuliert, was eine Anpassung an die Materialklasse ermöglicht. Der Einfluss dieser variablen Referenzsteifigkeit auf das effektive elastische und thermische Verhalten wird diskutiert und mit experimentellen Daten verglichen.

Weiterhin wird mittels einer Full-Field Methode, die auf der schnellen Fourier-Transformation (FFT) basiert, das lokale elastische und thermische Materialverhalten berechnet. Hierbei findet eine direkte Anwendung der Methode auf Voxel-Daten aus µCT Scans des Komposites statt. Die effektiven elastischen und thermischen Eigenschaften werden basierend auf den lokalen Feldern bestimmt. Somit können die Simulationsergebnisse der Full- und der Mean-Field Methode quantitativ bezüglich der Anisotropie des E-Moduls und des thermischen Ausdehnungskoeffizienten verglichen werden. Abschließend werden die numerischen Ergebnisse der effektiven elastischen und thermischen Materialparameter mit den experimentellen Daten verglichen und diskutiert.

Summary

Fiber-reinforced polymers contribute significantly to weight-reduced components in the automotive and aerospace applications. Due to the flowability of discontinuous fiber-reinforced polymers (DiCoFRP) during the manufacturing process, complex component geometries can be produced. In contrast to DiCoFRP, continuous fiber-reinforced polymers (CoFRP) achieve a high specific stiffness and strength due to their high fiber volume content. The aim of the research work within the International Research Training Group DFG-GRK 2078 is to combine the advantages of DiCoFRP and CoFRP. The resulting material class is characterized by the integration of functional elements into the manufacturing process, a high degree of design freedom for components and an economical mass production. The overall goal of the German-Canadian consortium of this Research Training Group is to develop concepts for the characterization, modeling, dimensioning and processing of DiCoFRP and CoFRP as well as the combination of both material systems.

In the work at hand, a discontinuous fiber-reinforced thermoset material is investigated produced by the Sheet Molding Compound (SMC) process. Due to the process-related fiber orientation distribution, a composite material with an anisotropic microstructure is created. This microstructure crucially affects the mechanical properties.

The central objectives of this work are, on the one hand, the modeling of the thermoelastic material behavior of the composite accounting for the underlying microstructure and, on the other hand, the experimental characterization of the pure resin and the composite material.

Dynamic mechanical analysis (DMA) is used to experimentally characterize the thermoviscoelastic material properties. Based on these tests, material-specific properties such as the temperature-dependent elastic modulus or the glass transition temperature are determined. In addition, measurements are carried out to determine the coefficient of thermal expansion of the pure matrix and composite material. The experimental data of the matrix material serve as input for the material models. The numerically determined material parameters are compared with the experimental data of the composite.

To approximate the effective thermoelastic material behavior, the Hashin-Shtrikman mean-field homogenization method is used which is based on the eigenstrain formulation by Willis (1981) and Walpole (1966). The effective quantities are computed in two steps. In this context, the homogenization method is formulated in dependency of a variable reference stiffness which enables an approach tailored to a specific material class. The influence of this variable reference stiffness on the effective elastic and thermal behavior is discussed, and compared to the experimental data.

Furthermore, the local elastic and thermal material behavior is computed using a full-field method based on the fast Fourier transformation (FFT). Here, the method is directly applied to voxel data obtained by μCT scans of the composite. The effective elastic and thermal properties are determined based on the local fields. Thus, the simulation results of the full- and mean-field methods can be compared quantitatively and with respect to the anisotropy of Young's modulus and the thermal expansion coefficients. Finally, the numerical results of the effective elastic and thermal material parameters are compared with the experimental data and discussions are provided.

Acknowledgments

First and foremost, I would like to express my sincere gratitude to Prof. Dr.-Ing. Thomas Böhlke for supervising my doctoral thesis with enthusiasm, guidance, and patience. His critical comments and insightful advices as well as his immense knowledge have been greatly supportive during the past four years. Encouraging me to participate in conferences gave me the opportunity to expand my knowledge and get new ideas for my research work.

I would also like to thank Prof. Jeff Wood for the co-supervision of my thesis, and the collaborative support during my research stay at the University of Western Ontario. I have benefited a lot from his expertise in experimental analyses.

I thank Prof. Dr. rer. nat. Britta Nestler for chairing the thesis committee.

Special thanks go to Jun.-Prof. Dr. rer. nat. Matti Schneider for fruitful discussions and sharing his huge knowledge on FFT and numerical homogenization. His comments and suggestions on microstructure segmentation and full-field modeling have been extremely enriching.

I greatly appreciated the collaborative work within the international research training group GRK 2078. During the long meetings, there was always a pleasant and constructive atmosphere. In particular, I would like to thank Pascal Pinter for the profound discussions on microstructure characterization, and for providing me the necessary input for my simulations. I also enjoyed the discussions and collaborative work with Anna Trauth, especially, during the manufacturing of the pure resin plaques.

The last years wouldn't have been so enjoyable without my colleagues and friends from the ITM-KM. During lunch time, coffee breaks and after work we always had interesting conversations on a variety of topics. In particular, I thank Daniel, Hannes, Johannes G., Jürgen, Malte, Mauricio, and Róbert for fruitful discussions. I also thank my colleagues from the ITM-Dynamics Ulrich, Tim, Olga, and Kai, for the enlightening and distracting conversations. Some special words of gratitude go to Helga and Ute, who were a great support in administrative issues and always had an open ear for private and professional concerns. Many thanks to Tom for his support in teaching and all Linux-related problems. I would also like to thank my student assistant Sebastian Reimann for his great support in the experiments over the last two years.

Very special thanks go to my friends Sara, Katharina K., Katharina G. and Maximilian for their friendship, patience and support.

Special gratitude is devoted to my sister Michelle, and to Max, Sigrun, Rudi and Katharina for their unfailing emotional support.

My heartfelt thanks go to my parents Michael and Emilia, who raised me, supported me, and never gave up on me. Their love and trust made me the person I am today.

I would like to express my appreciation and gratitude to my fiancé Andreas. I thank him for his support, love, and confidence in me. Especially, in the wide field of continuum mechanics, I could always count on him.

The research documented in this doctoral thesis has been funded by the German Research Foundation (DFG) within the International Research Training Group "Integrated engineering of continuous-discontinuous long fiber reinforced polymer structures" (GRK 2078). The support by the German Research Foundation (DFG) is gratefully acknowledged.

Karlsruhe, May 2019 Maria Loredana Kehrer

Contents

1 Introduction 1
1.1 Motivation 1
1.2 State of the Art 8
1.3 Outline of the Thesis 18
1.4 Notation, Frequently Used Acronyms, Symbols, and Operators 19

2 Fundamentals of Continuum Mechanical Material Theory . . 23
2.1 General Continuum Mechanical Equations 23
2.1.1 Kinematics 23
2.1.2 Balance of linear and angular momentum 24
2.1.3 Balance of internal energy 25
2.1.4 Dissipation inequality 26
2.2 Thermoelasticity with Temperature-Dependent Coefficients 27
2.2.1 Exploitation of dissipation inequality 27
2.2.2 Heat conduction 28
2.2.3 Constitutive modeling 29

3 Microstructure Characterization of Long Fiber-Reinforced Polymers 35
3.1 Fiber-Reinforced Composites 35
3.1.1 Classification of fiber-reinforced composites 35
3.1.2 Computed tomography scans 38
3.1.3 Mathematical description of a fibrous microstructure 39
3.2 Microstructure Characterization of SMC Composites . . . 42

4 Experimental Investigations of Pure and Reinforced Polymers 49
4.1 Dynamic Mechanical Analysis 49
4.1.1 Fundamentals 49
4.1.2 Experimental setup 54
4.1.3 Temperature- and frequency-dependent material behavior of pure and DiCo fiber-reinforced UPPH . . 57
4.1.4 Summary of results 67
4.2 Measurement of Thermal Expansion 68
4.2.1 Experimental setup 68
4.2.2 Validation based on aluminum 70
4.2.3 Thermal expansion and CTE of pure and DiCo fiber-reinforced UPPH 71
4.2.4 Summary of results 74

5 Homogenization for Fiber-Reinforced Polymers 75
5.1 Mean-Field Homogenization 75
5.1.1 Fundamental equations 75
5.1.2 Mean-field homogenization based on a variable reference stiffness 82
5.2 Full-Field Homogenization Based on Fast Fourier Transformation 85

6 Mean- and Full-Field Simulations for Long Fiber-Reinforced Polymers 89
6.1 Preliminaries 89
6.2 Mean-Field Homogenization for DiCo Fiber-Reinforced UPPH at $\theta = 20°\mathrm{C}$ 93
6.2.1 Effective elastic material properties 93
6.2.2 Summary of results 104
6.3 Full-Field Homogenization for DiCo Fiber-Reinforced UPPH at $\theta = 20°\mathrm{C}$ 105
6.3.1 Preliminary studies on synthetic fiber structures . . 105

6.3.2 Effective elastic material properties 108
6.3.3 Summary of results 113
6.4 Temperature-Dependent Effective Elastic Material Properties 114
6.4.1 Mean-field homogenization results 114
6.4.2 Summary of results 118
6.5 Effective Thermal Properties 118
6.5.1 Mean-field homogenization results for $\theta = 20°\mathrm{C}$ . . 118
6.5.2 Full-field homogenization results for $\theta = 20°\mathrm{C}$ . . . 121
6.5.3 Temperature-dependent effective thermal material properties 122
6.5.4 Summary of results 124

7 Summary and Conclusions 125

A Mean- and Full-Field Homogenization 131
A.1 Hill's Polarization Tensor 131
A.2 Hashin-Shtrikman Mean-Field Homogenization 133
A.2.1 Hashin-Shtrikman functional 133
A.2.2 Effective quantities 140
A.2.3 Mechanical strain and eigenstrain localization . . . 140
A.3 Orientation Tensors for the SMC Composite Samples . . . 143
A.4 Settings for the Full-Field Simulations 147

B Production of Pure Resin Samples 149

C Application to Sheet Molding Compound Composite 151
C.1 Mean-Field Simulation Results for 1D/2 Plaque 151
C.2 Mean-Field Simulation Results for $c_{\mathrm{F}} = 0.22$ 155

Bibliography 159

Chapter 1

Introduction

1.1 Motivation

A global superior goal of today's industrial companies is to comply with the restrictions on carbon dioxide emissions. In the European Union, for instance, car manufacturers are required to reduce their emissions values for all newly produced cars to $95\,\mathrm{g\,CO_2/km}$ by 2021 (European Union, 2018). One problem-solving approach is to substitute the non-structural metal parts with fiber-reinforced polymeric composite parts to reduce the fuel consumption. In addition to the weight savings, these materials exhibit certain advantages with respect to their material properties, such as exceptional specific strength and stiffness. As the production technology has progressed in the last decades, low cycle times have been achieved that enable a cost-effective manufacturing of fiber-reinforced polymers (FRP) up to mass production. Thus, various fields of applications have emerged, including not only the industrial sectors such as aerospace and transportation but also leisure and sports sectors, cf. Figure 1.1a. The tailored properties of composites are achieved by combining the polymeric matrix with a specific type of fiber. Thereby, the fibers carry the load; whereas, the matrix material defines the shape of the component, specifies the position of the fibers, and protects the fibers from environmental influences. Common types of reinforcements include glass, carbon, natural, and aramid fibers, cf. Figure 1.1b.

Generally, the fiber reinforcements are categorized into continuous (endless) and discontinuous (chopped) fibers. In terms of continuous fiber-reinforced polymers (CoFRP), a high fiber volume content of up to 60% can be achieved (Henning and Moeller, 2011). Typically, the fiber length exceeds 50 mm. Combined with a controlled fiber alignment, high stiffness and strength can be obtained with respect to the designated fiber direction. In contrast, the formability of CoFRP is limited, and long cycle times have to be considered. These drawbacks are overcome by using discontinuous fiber-reinforced polymers (DiCoFRP), albeit a lower stiffness and strength is given compared to CoFRP. In the context of DiCoFRP, a distinction between short and long fiber reinforcements is commonly made. Herein, short fibers exhibit fiber lengths of up to 1 mm, whereas long fibers have fiber lengths between 1 mm and 50 mm, (Ehrenstein, 1999).

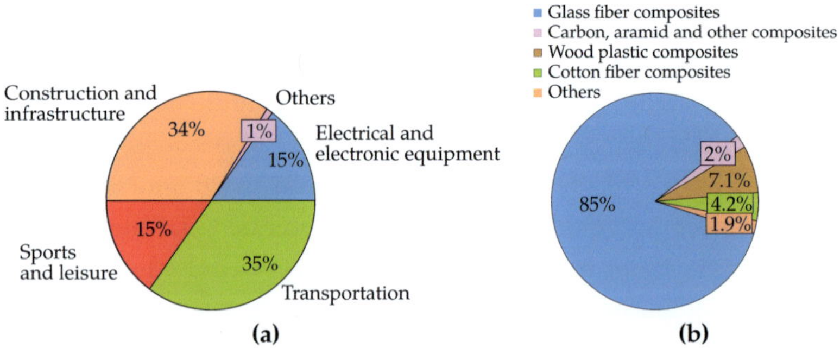

Figure 1.1: a) Distribution of glass fiber-reinforced composites by market segments in Europe in 2017 (Witten et al., 2017). b) Percentage of fiber types used in Europe in 2010 (Reddy, 2015; Carus, 2011)

Combining CoFRP and DiCoFRP to continuous-discontinuous fiber-reinforced polymers (CoDiCoFRP), a new material class is created. The International Research Training Group DFG-GRK 2018 aims to establish an integrated engineering of this new material class. Based on an interdisciplinary collaboration of the different projects, a virtual process chain is constructed to design the structural parts. The different projects within the process chain are illustrated in Figure 1.2.

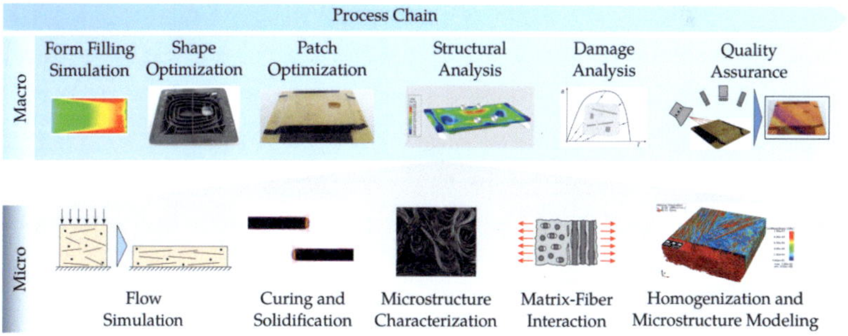

Figure 1.2: Virtual process chain consisting of partial steps. Simulations are performed on the macroscopic scale integrating results obtained at the microscopic scale.

A starting point in the process chain is given by the mold-filling simulation that focuses especially on the fiber orientation resulting from the manufacturing process, Hohberg et al. (2017). The local fiber orientation and the effective viscosity are analyzed on the microscopic level (Bertóti and Böhlke, 2017). During the production process, the curing and the solidification of the plastificate are initially described by a phase-field model on the microscopic scale (Schneider et al., 2017a). The microstructure of the cured composite material is characterized by means of μCT scans, by which the microstructural information such as the fiber orientation is obtained, cf. Pinter et al. (2018). Thus, the fiber-matrix interaction is investigated, and the interphase is characterized (Schober et al., 2017). Taking the foregoing findings into account, the effective material properties are determined by mean- and full-field simulations (Kehrer et al., 2018). The structural part on the macroscopic length scale is optimized by shape-optimization simulations (Albers et al., 2017). Areas exposed to particular high load are locally reinforced by patches of continuous fiber reinforcements. These areas and the resulting optimal patch positions are determined by patch optimization simulations, cf. Fengler et al. (2018). For specific load cases, a structural analysis is performed to investigate the damage behavior, cf. Schemmann et al. (2018b). Experimental tests are performed by Trauth et al. (2018) and serve to validate the foregoing simulations. Finally, the quality assessment for the semi-finished product, as well as the cured SMC composite, serves as an in-line process control as well as a quality control for the final composite part (Zaiß et al., 2017). A reference structural part, as shown in Figure 1.3, represents finally the successful implementation of the virtual process chain.

Figure 1.3: Reference structure within the DFG-GRK 2078/1 consisting of DiCo fiber-reinforced UPPH locally reinforced by patches of Co fiber reinforcements.

Research objectives of this thesis. Within the scope of the process chain, mainly three topics are addressed: the (I) mean- and (II) full-field homogenization to predict the effective thermoelastic material properties, and the (III) experimental characterization of the thermo-mechanical behavior using dynamic mechanical analysis (DMA) and thermal expansion measurements, cf. Figure 1.4. The material system considered is manufactured from sheet molding compound (SMC), and consists of a thermoset matrix reinforced with discontinuous (DiCo) glass fibers.

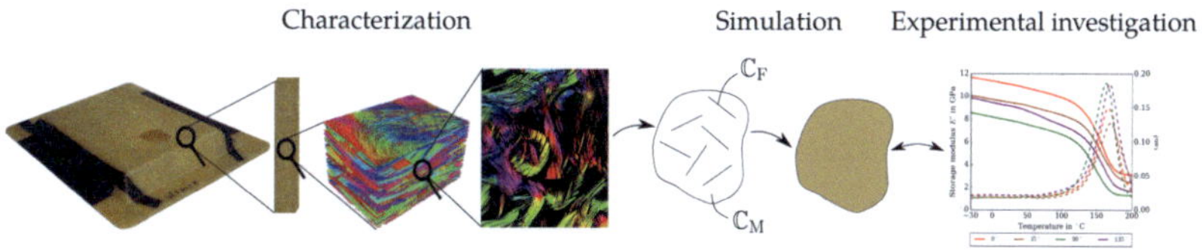

Figure 1.4: Reinforced samples of the same material system that was used for the reference structure are characterized in the first step. The microstructural information such as fiber orientation or length is obtained. These microstructure characteristics are taken into account in the simulation model, and the effective elastic properties are computed. The presented material behavior is compared to the experimental data determined.

In detail, the main objective (I) of this thesis is to develop a thermo-mechanical mean-field method that accounts for the fiber orientation distribution of the SMC composite and reliably predicts the effective elastic and thermal material properties. The orientation distribution is given in terms of orientation tensors obtained by micro-computed tomography (μCT) scans.

A benefit of the developed thermoelastic mean-field method is its numerically efficient formulation. Hence, it can be connected to the optimization or flow simulations and can thus be integrated into the process chain. This is especially important since some commercial software tools provide only simplified approaches for estimating the overall properties. These approaches are mainly based on a rule of mixture or on the Mori-Tanaka homogenization method. These built-in methods exhibit limitations when applied to anisotropic constituents or in terms of a multiphase composite with more than two distinct phases, for instance. Hence, considering a more complex microstructure requires a more elaborate method, such as the Hashin-Shtrikman (HS) two-step method presented in this thesis.

Furthermore, the HS method is formulated in terms of a variable reference stiffness which allows to tailor the mean-field model to a specific material class. This is beneficial in terms of composites with a high phase contrast.
The experimental analysis of the material reveal a strongly temperature-dependent elastic modulus and temperature-independent coefficient of thermal expansion (CTE) of the matrix. Thus, the HS two-step method is formulated in terms of a temperature-dependent stiffness to determine the effective elastic and thermal behavior not only at room temperature but also within a reasonable temperature range.

The second aim (II) is to obtain the overall material properties by application of a full-field method to raw μCT data. Based on the real data, the local material behavior is calculated by the fast Fourier transformation (FFT) method, and the effective properties are determined by volume averaging. Due to comparatively high calculation times, the full-field simulation is not directly applicable in combination with flow simulations, for instance. The effective anisotropic Young's modulus and CTE computed by the full-field simulation, however, can be compared to the results obtained by the mean-field simulation. This is necessary, since the experimental data are only given with respect to a specific point within the angular distribution of the effective quantities.
Furthermore, virtual microstructures that are related to real microstructures can be created for a profound analysis of the microstructural influence on the macroscopic material behavior. Due to synthetic microstructures, influences from the segmentation procedures are neglected.

The third objective (III) of this thesis is to experimentally characterize the SMC composite by means of DMA. The DMA system is a commonly used measuring technique to determine the thermoviscoelastic material parameters and the glass transition of polymers. The analysis of the material allows quantifying the admissible temperature range of possible applications. In addition, the thermal material behavior is characterized by thermal expansion measurements. The measured data are used to validate the numerical simulation results obtained by the mean- and full-field homogenization.

Originality of this thesis. The thermoelastic mean-field method developed exhibits the following novelties:

- **Direct use of orientation tensors:** Many mean-field methods only account for the microstructure via orientation distribution functions. These, in turn, are derived from orientation tensors by taking additional assumptions into account. To overcome this shortcoming, an approach is developed that directly uses orientation tensors. The benefit of this approach is that it enables a simple application in combination with filling or flow simulations using, e.g., Moldflow®. Methods, such as the Mori-Tanka estimate, are also formulated in terms of orientation averages but can yield non-physical results when applied to anisotropic constituents, for instance. Since the approach developed does not exhibit such limitations, it is generally applicable and particularly suitable for CoDiCo composites.
- **Simple mathematical structure:** The effective elastic stiffness approximated by a Hashin-Shtrikman (HS) approach is formulated as stated in Walpole (1966). This form is mathematical identical with the expression given in Willis (1977; 1981) but offers the advantage of a simpler structure regarding the averaging performed on the tensors. Since, the averaging of the tensors needs to be performed only once, instead of twice, computation efforts are reduced, cf. Lobos Fernández (2018).
- **Variable reference stiffness:** The HS mean-field method is introduced in terms of a variable reference stiffness which allows a tailored approach to determine the effective thermoelastic material properties. Since the choice of the proper reference material is not

always unambiguous, a variable chosen reference stiffness allows accounting for characteristics of a specific material class. Thereby, the material modeling can be adjusted to a specific composite.

- **Temperature-dependent stiffness:** The HS two-step method is formulated in terms of a temperature-dependent stiffness. This enables to predict the effective elastic and thermal properties for the entire temperature range in which the composite can be used.

The experimental investigations by means of dynamic mechanical analysis (DMA) and measurements of the thermal expansion with a video extensometer system provide the following new insights:

- **New material class:** A new material system consisting of a fiber-reinforced polyester-polyurethane hybrid resin is considered. Thus, an experimental characterization of the elastic and thermal properties provides information about the material behavior. As a consequence, conclusions can be drawn with respect to the material modeling of this new material class.
- **Field of application:** Using DMA, the limitation of the field of application, such as the temperature range of a possible application, is identified by the glass transition.

1.2 State of the Art

Determining the microstructure. When using micromechanical approaches, profound information on the microstructure, including fiber length, orientation, and volume fraction, is of crucial importance. During the last decades, various methods for characterizing the microstructure of a material have been developed. Generally, a classification into destructive and non-destructive methods can be made. An often-applied destructive method is represented by the 2D elliptical section method. Applying this method, the sample is cut by a microtome, and the surface is polished. On the polished surface, the elliptical footprints of the fibers are visible and can be subsequently evaluated by various image analysis software tools. Herein, the angle between the fiber axis and the direction perpendicular to the cutting surface can be determined (Fakirov and Fakirova, 1985). Although this method is destructive, it is still used due to its simplicity and low cost. Hine et al. (2014) applied this method to determine the fiber orientation distribution and compute the effective elastic tensile properties using a modified rule of mixture. Major errors of the sectioning method are expected in terms of fibers that are perpendicular to the cutting surface and thus exhibit a circular cross-section (Clarke et al., 1993). Hence, this method works best in terms of planar-oriented fibers.

Considering the non-destructive characterization of composites, the computed tomography (CT) scans have gained significance, cf. Maire and Withers (2014). In terms of µCT methods, the X-rays intersect a sample which rotates around its center axis. Depending on the absorption capacity of the constituents of the composite, the radiation hits the detector plane behind the sample and produces a radiograph. Some limitations arise in terms of the materials with a low absorption index, such as carbon fibers in a polymer matrix. This drawback can be overcome by applying non-conventional X-ray sources such as synchrotron light (Bernasconi et al., 2008). From the projections, the internal 3D microstructure can be reconstructed by stacking the 2D images obtained previously (Feldkamp et al., 1984). The 3D voxel-based image obtained is characterized by gray values and can be processed for analyzing the internal fiber architecture such as the fiber orientation distribution. In this context, many methods can be found in the literature, cf. Krause et al. (2010); Ohser and Schladitz (2009); Bernasconi et al. (2012). In terms of low fiber volume contents, every single fiber can be isolated, and the orientation angles are determined (Shen et al., 2004). For increasing fiber volume contents, other methods are applied that are based on specific filtering algorithms applied to each voxel of a reconstructed μCT scan. The overall fiber orientation within a voxel-based image can be determined by means of common methods like the Gaussian filter (Robb et al., 2007) or the Hessian matrix (Daniels et al., 2006).

The microstructure analyzing tools or simulation methods are often applied with respect to short fiber-reinforced composite. However, some applications are given for injection-molded long fiber-reinforced plastics, cf. Denault et al. (1989); Hine et al. (1995); Phelps and Tucker III (2009).

A general and complete description of the fiber orientation is given by the orientation distribution function describing the probability of finding a fiber of a specific orientation (Bay and Tucker III, 1992). This might become time-consuming when computing more complex materials. Introduced by Kanatani (1984), the mean fiber orientation distribution can be quantitatively described in terms of fiber orientation tensors. This representation is popular, especially in the context of simulating flow-induced fiber orientation during the manufacturing process. For injection molded applications, Advani and Tucker III (1987) presented methods to compute the fiber orientation tensor of the fourth-order based on the second-order tensor using a linear, quadratic, and hybrid closure approximation.

In recent research works, the fiber length distribution is considered in the material models, additionally to the fiber orientation distribution. The studies and investigations performed aim at determining the influence of the fiber length distribution and the volume fraction on the mechanical properties, cf. Hine et al. (2002); Nguyen et al. (2008); Garesci and Fliegener (2013).

Generally, for predicting the overall material behavior, both the raw data obtained directly by conventional or synchrotron μCT scans and the extracted orientation information given by orientation distributions or orientation tensors are commonly used in modern methods. Müller et al.

(2016) computed the effective elastic properties by means of orientation tensors combined with mean-field homogenization. An extension for considering the damage behavior is given in Schemmann (2018). Direct use of the imported raw data is made using full-field methods such as the Finite Element method (FEM) or the fast Fourier transformation (FFT) method, cf. Schneider et al. (2016); Düster et al. (2012); Węglewski et al. (2013). But also, orientation tensors are used combined with an FFT-based homogenization, as presented in Goldberg et al. (2017), for computing the effective nonlinear material properties.

Dynamic mechanical analysis. Dynamic mechanical analysis (DMA) has become a widely used technique to characterize the thermoviscoelastic behavior of materials. Using DMA, a stress or strain load is applied to a sample, and the material's properties obtained are analyzed as a function of time, temperature, frequency, or load (Menard, 2008). Since the technique is highly sensitive with respect to the movements of the polymer chains, it represents a tool ideally suited for measuring the transitions in polymers. Common synonyms throughout the literature for DMA are dynamic mechanical thermal analysis (DMTA), dynamic mechanical spectroscopy, or dynamic thermomechanical analysis.

One of the first attempt to measure the material's response to an oscillating load was made by Poynting et al. (1909). The first commercial instruments for rheological measurements were developed in the 1950s, such as the Weissenberg rheogoniometer or the Rheovibron (Gabbott, 2008). The viscoelastic properties were determined by cone-to-plate measurements. In his first edition from 1961, Ferry (1980) presented the phenomenological theory of linear viscoelasticity combined with measured viscosity data. Since 1966, the modern period of DMA began with the introduction of the torsional analyzer by Gillham, cf. Gillham and Enns (1994). The early DMA systems, however, were difficult to use and exhibited limitations with respect to the processability of the data. In the early 1980s, Polymer Labs launched a DMTA for axial sample geometries. Following the instruments by Du Pont, PerkinElmer and Triton Technologies developed sensors with higher sensitivity and easy-to-use designs (Gabbott, 2008). The instruments developed offered measurements with loading of up to maximal 40 N (Mettler Toledo), whereas GABO developed systems with a loading of up to 150 N by 1981 and even 1500 N by 1988. Since the 1980s, the instruments have become more user-friendly due to the progress in computers and software.

DMA systems have been applied to various composite systems. Some attempts have been made by Reed (1980) and Mijović and Liang (1984) to detect the presence of interphases in unidirectional carbon fiber (CF)-reinforced polymers by applying DMA. Investigations by Thomason (1990) for both glass and carbon reinforced epoxy composites revealed that the interphases cannot be clearly detected due to the complex interaction between matrix and fibers. Akay (1993) applied DMA to a wide range of polymer systems, including short and long glass fiber (GF)-reinforced thermoplastics as well as unidirectional CF epoxy. Especially in the context of short fiber-reinforced polymers, studies on the viscoelastic material properties were carried out by Tjong et al. (2002). The dependency of the elastic and viscous material parameters on the fiber volume content was investigated by Fung and Li (2006). Mouzakis et al. (2008) used an environmental aging chamber and a DMA system to experimentally investigate the failure of polymers exposed to environmental effects. Their studies revealed a stiffening effect of the matrix material exposed to UV radiation, accompanied by microcracks. The behavior of a pre-damaged unidirectional CF-reinforced epoxy is characterized in Kostka et al. (2016). In the context of a long GF-reinforced thermoplastic, the nonlinear time-, temperature-, and frequency-dependent behaviors were determined by Brylka et al. (2018), who also derived the material's parameters for the effective elastic properties.

Mean-field homogenization. Since materials exhibit a heterogeneous microstructure at different length scales, the prediction of the overall properties accounting for the underlying heterogeneities is the subject of current and past research. Generally, the existing methods for predicting the macroscopic material behavior can be divided into the bounding and estimating methods (Dvorak, 2012).

Modern mean-field methods can be traced back to approaches formulated by Voigt (1889) and Reuss (1929). Based on the principle of minimum potential and complementary energy, the effective stiffness and compliance, respectively, were derived for predicting the linear elastic behavior of a polycrystal. Hill (1952) showed that the approach by Voigt represents the upper bounds, whereas Reuss formulated the lower bounds for the linear elastic material properties. Since both methods only account for the volume fraction as microstructural information, these bounds are denoted as first-order bounds. Hashin and Shtrikman (1962) derived closer bounds that are based on a variational principle, cf. also Walpole (1966; 1969). These bounds consider additional information on the geometry of the inclusions and thus represent second-order bounds. A generalized formulation accounting for eigenstresses and eigenstrains is derived by Willis (1977; 1981), introducing Green's function for a homogeneous comparison material. More elaborate and higher-order bounds were introduced by Kröner (1977) and applied to particulates and fibrous materials in Milton (1985; 2002) and Torquato (2002). In terms of anisotropic polycrystalline materials, enhanced Hashin-Shtrikman bounds were developed by Böhlke and Lobos (2014). Initiated by Willis (1983), the Hashin-Shtrikman variational principle was extended to account for the nonlinear material behavior of composites, cf. Talbot and Willis (1985). This attempt was generalized by Castañeda (1991), introducing a heterogeneous comparison material to determine the bounds for nonlinear behavior. These second-order bounds were, for instance, applied to porous media (Idiart and Castañeda, 2005) or for determining the viscoplastic behavior of polycrystals (Liu and Castañeda, 2004).

Based on Eshelby's solution for a single inclusions problem (Eshelby, 1957), various approximation methods for estimating the effective material properties have been developed. One of the simplest approximation methods is given by the dilute distribution method, cf. Torquato (2002). Herein, the inclusions or particles do not interact with each other but are dilute distributed. Based on these assumptions, Maxwell (1891) developed the effective conductivity, and Einstein (1906) the effective shear viscosity of spheres dispersed in a fluid. One drawback of this method is given in terms of higher concentrations of the inclusions so that the Voigt and Reuss bounds might be violated (Nemat-Nasser and Hori, 1993).

This drawback is overcome by the approximation method proposed by Mori and Tanaka (1973). Using this method, the inclusions are embedded into an infinite matrix subjected to an effective stress field which equals the average stress of the matrix material. Due to the simple structure of the Mori-Tanaka method, there are numerous applications, e.g., Benveniste et al. (1989); Chen et al. (1990); Zhao et al. (1989). Benveniste (1987) has shown that in more general cases, such as for multiphase composites with different aligned or shaped inclusions, the fundamental condition of a symmetric effective stiffness is not necessarily ensured. For the special case of isotropic constituents and unidirectional fibers, however, Tucker III and Liang (1999) showed that the effective elastic properties obtained by the Mori-Tanaka method coincide with the results obtained by the lower Hashin-Shtrikman bounds. In Brassart et al. (2009), an extended formulation of the Mori-Tanaka method accounts for nonlinear constituents subjected to finite strains. The method applied has limitations when a failure mechanism occurs. Recent applications of an extended Mori-Tanaka method, formulated in terms of orientation tensors, are given by Brylka (2017). An approach to characterize the heterogeneous microstructure of sheet molding compound composites by means of biaxial testing and an inverse parameter identification is presented in Schemmann et al. (2018a;c). In combination with an enhanced Mori-Tanaka mean-field model, the damage behavior is investigated (Schemmann et al., 2018b).

Another well-known approximation method is given by the self-consistent scheme, introduced by Hershey (1954) and Kröner (1958). For this purpose, an inclusion is embedded in an infinite matrix with the properties of the effective, unknown material. Despite its implicit mathematical structure, the self-consistent scheme is widely applied, cf. Budiansky (1965); Hill (1965). The thermoelastic material properties were derived by Laws (1973) and extended to linear viscoelastic materials in Laws and McLaughlin (1978). Various methods were derived based on the self-consistent scheme. The generalized self-consistent scheme by Christensen and Lo (1979) is a well-known representative. Related to the three-phase model, the heterogeneity is embedded into a matrix which, in turn, is surrounded by the effective medium. The generalized self-consistent method exhibits restrictions to spherical or cylindrical heterogeneities (Benveniste, 2008). Aiming for a general use, Zheng and Du (2001) developed the effective self-consistent scheme, which is valid for multiphase composites with arbitrarily shaped matrix-inclusion cells. A major shortcoming of the latter method is its complicated structure. Thus, linearizing the effective self-consistent method results in the interaction direct derivative method (Zheng and Du, 2001; Du and Zheng, 2002).

In recent publications, use is made of two-step homogenization methods, cf. Pierard et al. (2004). In this context, the considered domain is divided into subdomains consisting of two phases, a matrix and an inclusion phase (Lielens et al., 1998; Camacho et al., 1990). In the first step, the homogenized solution of each subdomain is determined. The overall properties are obtained by averaging over all subdomains. Since a subdomain consists of only a two-phase material, the Mori-Tanaka method is applicable and thus often used, cf. Doghri and Tinel (2005). For the second step, the Voigt or Reuss methods are applied (Doghri and Friebel, 2005). In Müller (2016), a two-step method applied to a short fiber-reinforced composite is compared to the self-consistent scheme and the interaction direct derivative. n-step or hierarchical homogenization methods have been developed especially in the context of hybrid composite systems, cf. Jendli et al. (2009). Anagnostou et al. (2018) applied a hierarchical homogenization method to a SMC consisting of three homogenization steps at different length scales. The homogenization methods used in the corresponding steps are the Mori-Tanaka method and the composite cylinders method (Christensen and Lo, 1979; Meraghni and Benzeggagh, 1995; Matzenmiller and Gerlach, 2004). Thereby, the effective viscoelastic behavior is determined, as well as the anisotropic damage behavior, considering a penny-shaped microcrack density of the matrix.

Full-field homogenization. In recent years, the considered composite systems and applications exhibited increasingly complex microstructures. In this context, full-field methods, such as the finite element method (FEM) or the fast Fourier transformation (FFT) approach, are powerful tools to analyze the local and overall behavior of multiphase composites. Generally, in the context of full-field homogenization, a boundary value problem for a representative volume element or a unit cell is defined. Caused by external loads, the local solution is obtained by applying the full-field method. The homogenized properties can be computed by volume averaging.

Using the FEM, the viscous and damage behavior of a long fiber-reinforced thermoplastic is investigated by Fliegener (2015). Numerical damage models with single and multi-fibers are considered in Wang et al. (2009). Applications to linear as well as nonlinear problems in the context of viscoelasticity and thermoelasticity are presented in Ostoja-Starzewski (2006).

Considering the FFT-based methods, no meshing or assembling is necessary, and thus, arbitrary and complex microstructures can be considered. Furthermore, the FFT-based computations are much faster than computation by applying the FEM, for instance, cf. Michel et al. (1999). Moulinec and Suquet (1994; 1998) first proposed an iterative FFT method to solve the integral Lippmann-Schwinger equations (Kröner, 1977) that are equivalent to the periodic boundary value problem in linear elasticity. In recent years, attempts have been made to speed up solvers or to extend the field of applications and to account for geometrically and physically nonlinear material behavior. In Michel et al. (2001), modified methods are derived for composites with high or infinite contrasts between the constituents of a composite. These methods allow for the handling of composite structures containing voids, rigid inclusions, or exhibiting nonlinear material behavior, cf. also Monchiet and Bonnet (2013). Vinogradov and Milton (2008) proposed applying an enhanced FFT method to compute the thermoelastic material properties. Various methods have been proposed to solve nonlinear Lippmann-Schwinger equations and to determine the material behavior at finite strains (Lahellec et al., 2003; Eisenlohr et al., 2013). Algorithms for investigating damage phenomena (Sliseris et al., 2014) and progressive damage in composite materials have been developed (Spahn et al., 2014).

Kabel et al. (2014) developed a numerically efficient Newton-Krylov solver to compute the FFT-based homogenized properties at large deformations. The significant reduction in the computational effort was exemplarily demonstrated on a two-phase laminate and a glass fiber-reinforced polymer structure. Schneider (2017a) developed a gradient-based solver that can be directly applied to inelastic problems.

A comprehensive benchmark study is performed by Andrä et al. (2013) for various segmented 3D images of geomaterials. In this study, the effective physical properties like the electrical conductivity or elastic modulus are computed by various full-field methods such as the finite difference method, FFT, or FEM. Thus, conclusions are drawn with respect to the solver type, the boundary condition, or the segmentation process.

Discretization by trigonometric polynomials leads to an ill-conditioned problem in terms of high material contrasts. Schneider et al. (2016) proposed a discretization method by finite differences on a staggered grid to obtain a fast and robust solver for linear elastic problems. A generalization by trilinear hexahedral elements for physically nonlinear material behavior is presented in Schneider et al. (2017b). Kabel et al. (2015) introduced a discretization method to treat interface voxels. This so-called composite voxel technique increases the accuracy of the computations with respect to local and effective elastic quantities. An extension to the inelastic material behavior of composites is given in Kabel et al. (2017).

1.3 Outline of the Thesis

The structure of this thesis is as follows.

Chapter 2 presents the continuum mechanical fundamentals and, particularly, the thermoelastic constitutive equations with temperature-dependent coefficients.

Chapter 3 addresses the characterization of the microstructure of the SMC composite considered in this thesis. First, a general classification of the fiber-reinforced composites is provided. Subsequently, the microstructure of a DiCo fiber-reinforced composite sample is analyzed based on reconstructed μCT data. This analysis reveals the characteristic properties of the material considered.

Chapter 4 introduces basic information with respect to DMA and, particularly, to the measuring system used. Experimental investigations of the pure and fiber-reinforced thermoset material are discussed. Moreover, the experimental setup for the measurements of the thermal expansion coefficients is presented, and the results obtained are discussed.

Chapter 5 presents the mean- and full-field methods for predicting the elastic and thermal properties. In terms of the mean-field method, a variable reference stiffness is introduced, and the effective stiffness and thermal expansion are determined by means of orientation averaging. Moreover, the fundamentals of the FFT are explained for the full-field method.

In Chapter 6, the application of the homogenization methods is presented with respect to the SMC composite. The predicted elastic and thermal properties are compared to experimental data from the DMA and CTE measurements, and the results observed are discussed.

Finally, in Chapter 7, concluding and summarizing remarks are given.

1.4 Notation, Frequently Used Acronyms, Symbols, and Operators

In this thesis, a direct tensor notation is preferred. Tensor components are indicated by Latin subscripts and Einstein's summation convention is applied. Frequently used symbols, acronyms and operators are listed below.

Acronyms

µCT	Micro-computed tomography
CF	Carbon fiber-reinforced
CoDiCoFRP	Continuous-discontinuous fiber-reinforced polymer

CoFRP	Continuous fiber-reinforced polymer
CTE	Coefficient of thermal expansion
DiCoFRP	Discontinuous fiber-reinforced polymer
FEM	Finite element method
FFT	Fast Fourier transformation
GF	Glass fiber-reinforced
HS	Hashin-Shtrikman
HS^{+}	Upper Hashin-Shtrikman bound
HS^{-}	Lower Hashin-Shtrikman bound
LFRP	Long fiber-reinforced polymer
ODF	(Fiber) orientation distribution function
RVE	Representative volume element
SMC	Sheet molding compound
UD	Unidirectional (microstructure type)
UPPH	Unsaturated polyester-polyurethane hybrid resin
vol.%	Volume fraction in %

Latin letters

$a, b, A, B, \mathcal{D}, \ldots$	Scalar quantities
$\boldsymbol{u}, \boldsymbol{v}, \boldsymbol{w}, \ldots$	First-order tensors
$\boldsymbol{A}, \boldsymbol{B}, \boldsymbol{C}, \ldots$	Second-order tensors
$\mathbb{A}, \mathbb{B}, \mathbb{C}, \ldots$	Fourth-order tensors
E'	Storage modulus
E''	Loss modulus
E	Young's modulus
c_{F}	Fiber volume fraction
c_{M}	Matrix volume fraction
$f(\boldsymbol{n})$	Fiber orientation distribution function
l_{F}	Fiber length
v	Volume
$\boldsymbol{a}$	Second-order eigenstrain localization tensor

$\boldsymbol{G}$	Eigenstress tensor
$\boldsymbol{I}$	Second-order identity tensor
$\boldsymbol{N}$	Orientation tensor of second-order
$\mathbb{A}$	Strain localization tensor
$\mathbb{B}$	Stress localization tensor
$\mathbb{C}_0$	Stiffness tensor of a reference medium
$\mathbb{C}$	Stiffness tensor
$\mathbb{G}_0$	Periodic Green operator for isotropic reference medium
$\mathbb{N}$	Orientation tensor of fourth-order
$\mathbb{P}$	Hill's polarization tensor
$\mathbb{I}^{\mathsf{S}}$	Symmetric fourth-order identity tensor
$\mathbb{P}_1$	First isotropic projector: $\mathbb{P}_1 = \boldsymbol{I} \otimes \boldsymbol{I}/3$
$\mathbb{P}_2$	Second isotropic projector: $\mathbb{P}_2 = \mathbb{I}^{\mathsf{S}} - \mathbb{P}_1$

Greek letters

α_θ	Coefficient of thermal expansion
ν	Poisson's ratio
$\tan\delta$	Loss factor: $\tan\delta = E''/E'$
θ	Temperature
$\boldsymbol{\alpha}_\theta$	Thermal expansion tensor
$\boldsymbol{\varepsilon}$	Infinitesimal strain tensor
$\boldsymbol{\varepsilon}^*$	Eigenstrain tensor
$\boldsymbol{\varepsilon}_\theta$	Thermal strain tensor
$\boldsymbol{\sigma}$	Cauchy stress tensor

Operators

$\boldsymbol{AB}$	Linear mapping of a second-order tensor
$\boldsymbol{A} = \mathbb{C}[\boldsymbol{B}]$	Linear mapping of a second-order tensor by a fourth-order tensor

$\boldsymbol{A} \cdot \boldsymbol{B}$	Dot product of two tensors $\boldsymbol{A}, \boldsymbol{B}$
$\boldsymbol{A} \otimes \boldsymbol{B}$	Dyadic product of two tensors $\boldsymbol{A}, \boldsymbol{B}$
$\mathbb{C} = \boldsymbol{A} \square \boldsymbol{B}$	Box product: $(\boldsymbol{A} \square \boldsymbol{B})[\boldsymbol{C}] = \boldsymbol{A}\boldsymbol{C}\boldsymbol{B}$
$\det(\cdot)$	Determinant
$\operatorname{div}(\cdot)$	Eulerean divergence of a tensorial quantity of second order and higher
$\operatorname{Grad}(\cdot)$	Lagrangian gradient of a quantity
$\operatorname{grad}(\cdot)$	Eulerean gradient of a quantity
$\operatorname{sym}(\cdot)$	Symmetric part of a quantity
$\overline{(\cdot)}$	Effective (macroscopic) quantity
$\langle\cdot\rangle$	Volume/ensemble average of a quantity
$(\cdot)_\alpha$	Quantity with respect to phase α
$(\cdot)_0$	Reference quantity
$(\cdot)^{\mathsf{T_H}}$	Major transposition of a fourth-order tensor, $C^{\mathsf{T_H}}_{ijkl} = C_{klij}$
$(\cdot)^{\mathsf{T_L}}$	Left minor transposition of a fourth-order tensor, $C^{\mathsf{T_L}}_{ijkl} = C_{ijlk}$
$(\cdot)^{\mathsf{T_R}}$	Right minor transposition of a fourth-order tensor, $C^{\mathsf{T_R}}_{ijkl} = C_{jikl}$
$(\cdot)^{\mathsf{T_M}}$	Mid minor transposition of a fourth-order tensor, $C^{\mathsf{T_M}}_{ijkl} = C_{ikjl}$

Chapter 2

Fundamentals of Continuum Mechanical Material Theory

2.1 General Continuum Mechanical Equations

2.1.1 Kinematics

In continuum mechanics, a body is defined as a set of material points. Regarding the Cauchy continuum, each material point has three translational degrees of freedom. A deformation χ describes the change of the geometry of the body due to kinematic boundary conditions or loads applied. In this context, it describes the shape of the body in a current configuration with respect to an arbitrarily chosen reference configuration. Regarding the reference configuration, the position of a material point is identified by a vector $\boldsymbol{X}$. In the current configuration, the position of the same material point is defined by the vector

$$\boldsymbol{x} = \chi(\boldsymbol{X}, t). \tag{2.1}$$

In order to describe the deformation of a continuum, constraints have to be introduced concerning $\chi(\boldsymbol{X}, t)$. Thus, $\chi(\boldsymbol{X}, t)$ has to be invertible, and continuously and twice continuously differentiable, cf. Bertram (2005). In this context, cracks or damaged material is not considered. The velocity of the body is given by the time derivative of the current position

$$v(\boldsymbol{X}, t) = \dot{\boldsymbol{x}} = \frac{\partial \boldsymbol{\chi}(\boldsymbol{X}, t)}{\partial t}|_{\boldsymbol{X}=\text{const.}}. \tag{2.2}$$

The change of a line element $\mathrm{d}\boldsymbol{X}$ of the reference configuration, due to the deformation, is described by a two-point tensor $\boldsymbol{F}$. Thus, the line element $\mathrm{d}\boldsymbol{x}$ of the current configuration is obtained by $\mathrm{d}\boldsymbol{x} = \boldsymbol{F}\,\mathrm{d}\boldsymbol{X}$. The tensor $\boldsymbol{F}$ is given by means of the gradient of the deformation $\boldsymbol{\chi}(\boldsymbol{X}, t)$ with respect to $\boldsymbol{X}$. Therefore, $\boldsymbol{F}$ is denoted as the deformation gradient and is defined by

$$\boldsymbol{F} = \mathrm{Grad}\,(\boldsymbol{\chi}(\boldsymbol{X}, t)) = \frac{\partial \boldsymbol{\chi}(\boldsymbol{X}, t)}{\partial \boldsymbol{X}}, \quad \boldsymbol{F} \in \mathrm{Inv}^+. \tag{2.3}$$

To describe the displacement of material points, the displacement field $\boldsymbol{u}(\boldsymbol{X}, t) = \boldsymbol{\chi}(\boldsymbol{X}, t) - \boldsymbol{X}$ is used. Its gradient with respect to $\boldsymbol{X}$ is denoted as the displacement gradient $\boldsymbol{H}(\boldsymbol{X}, t)$ and is given by

$$\boldsymbol{H}(\boldsymbol{X}, t) = \mathrm{Grad}\,(\boldsymbol{u}(\boldsymbol{X}, t)) = \frac{\partial \boldsymbol{u}(\boldsymbol{X}, t)}{\partial \boldsymbol{X}}. \tag{2.4}$$

2.1.2 Balance of linear and angular momentum

Balance equations relate the rate of change of an additive quantity in a considered volume to the production and supply of this quantity in the corresponding volume and the associated non-convective flux across the boundary (Müller, 1985). Subsequently, a material volume v is considered. Applying a localization procedure to the integral form of a balance equation yields a local form of the balance equation.

In classical continuum mechanics, five balance equations hold, independently of the material properties, and serve as universal physical laws (Truesdell and Toupin, 1960). These are the balances of mass, linear momentum, angular momentum, energy, and entropy. For a given mass density ρ, in the following, use is made of the local form of the balance of mass, i.e. $\dot{\rho} + \rho \mathrm{div}\,(\boldsymbol{v}) = 0$, with $\dot{\rho} = \partial\rho/\partial t + \boldsymbol{v} \cdot \mathrm{grad}\,(\rho)$. The rate of change of the linear momentum $\rho \boldsymbol{v}$ is related to the resultant external forces, consisting of surface and volume specific forces, $\boldsymbol{t}$ and $\rho \boldsymbol{b}$, respectively,

$$\frac{\mathrm{d}}{\mathrm{d}t} \int_v \rho \boldsymbol{v} \,\mathrm{d}v = \int_v \rho \boldsymbol{b} \,\mathrm{d}v + \int_{\partial v} \boldsymbol{t} \,\mathrm{d}a. \tag{2.5}$$

The surface traction field $\boldsymbol{t}$ on ∂v is related to the Cauchy stress $\boldsymbol{\sigma}$ by the Lemma of Cauchy, i.e. $\boldsymbol{t} = \boldsymbol{\sigma n}$, with the outward normal $\boldsymbol{n}$. By means of the divergence theorem and the Reynolds transport theorem for regular volumes, and considering the balance of mass, the local form of the balance of linear momentum is given by

$$\rho \dot{\boldsymbol{v}} = \rho \boldsymbol{b} + \mathrm{div}\left(\boldsymbol{\sigma}\right), \tag{2.6}$$

which reduces to $\boldsymbol{0} = \rho \boldsymbol{b} + \mathrm{div}\left(\boldsymbol{\sigma}\right)$ in the case of a quasi-static equilibrium. The balance of angular momentum in the integral formulation reads

$$\frac{\mathrm{d}}{\mathrm{d}t} \int_v \boldsymbol{x} \times \rho \boldsymbol{v} \,\mathrm{d}v = \int_v \boldsymbol{x} \times \rho \boldsymbol{b} \,\mathrm{d}v + \int_{\partial v} \boldsymbol{x} \times \boldsymbol{t} \,\mathrm{d}a, \tag{2.7}$$

resulting to Boltzmann's axiom, stating the symmetry of the stress tensor (Bertram, 2005)

$$\boldsymbol{\sigma} = \boldsymbol{\sigma}^{\mathsf{T}}. \tag{2.8}$$

2.1.3 Balance of internal energy

The first law of thermodynamics relates the rate of the total energy to the power expended by surface and volume forces as well as by the heat supply in the volume and the heat flow across the boundary of the volume. Its integral form is given by

$$\frac{\mathrm{d}}{\mathrm{d}t}\int_v \rho e + \frac{1}{2}\rho \boldsymbol{v}\cdot\boldsymbol{v}\,\mathrm{d}v = \int_v \rho \boldsymbol{b}\cdot\boldsymbol{v} + \rho r\,\mathrm{d}v + \int_{\partial v} \boldsymbol{t}\cdot\boldsymbol{v} - \boldsymbol{q}\cdot\boldsymbol{n}\,\mathrm{d}a, \quad (2.9)$$

where e represents the internal energy, $\boldsymbol{q}$ the heat flux vector, and r the heat supply. Taking into account the balance of mass and linear momentum, the balance of internal energy is obtained

$$\rho\dot{e} = \boldsymbol{\sigma}\cdot\boldsymbol{D} - \operatorname{div}(\boldsymbol{q}) + \rho r. \quad (2.10)$$

The rate of deformation tensor $\boldsymbol{D}$ is the symmetric part of the velocity gradient $\boldsymbol{L} = \dot{\boldsymbol{F}}\boldsymbol{F}^{-1}$, i.e. $\boldsymbol{D} = (\boldsymbol{L} + \boldsymbol{L}^{\mathsf{T}})/2$.

2.1.4 Dissipation inequality

The balance of entropy reads

$$\frac{\mathrm{d}}{\mathrm{d}t}\int_v \rho\eta\,\mathrm{d}v = \int_v \rho r_\eta\,\mathrm{d}v + \int_v \rho\gamma\,\mathrm{d}v - \int_{\partial v} \boldsymbol{q}_\eta\cdot\boldsymbol{n}\,\mathrm{d}a, \quad (2.11)$$

where η denotes the entropy, γ the entropy production, $\boldsymbol{q}_\eta$ the entropy flux, and r_η the entropy supply. According to the second law of thermodynamics, the entropy production is non-negative, i.e. $\Gamma = \int_v \rho\gamma\,\mathrm{d}v \geq 0$ for all thermodynamical processes. For a reversible thermodynamical process, the entropy production vanishes, i.e. $\Gamma = 0$, while $\Gamma > 0$ holds for an irreversible thermodynamical process. The local form of the entropy balance can be transformed into the local dissipation inequality

$$\rho\delta = \rho\theta\dot{\eta} - \rho\theta r_\eta + \theta\operatorname{div}(\boldsymbol{q}_\eta) \geq 0, \quad (2.12)$$

where $\delta = \gamma\theta$ is referred to as dissipation. For the entropy supply and the entropy flux, $r_\eta = r/\theta$ and $\boldsymbol{q}_\eta = \boldsymbol{q}/\theta$ are assumed (Coleman and Noll, 1963). The internal energy can be expressed by means of the Legendre transformation (Beegle et al., 1974; Coleman and Gurtin, 1967)

$$e = \theta\eta + \psi, \tag{2.13}$$

with ψ as the Helmholtz free energy density. Considering the Legendre transformation and the local form of the balance of internal energy, the local dissipation inequality reduces to the Clausius-Duhem inequality, reading

$$\rho\delta = \boldsymbol{\sigma} \cdot \boldsymbol{D} - \rho\dot{\psi} - \rho\dot{\theta}\eta - \frac{1}{\theta}\boldsymbol{q} \cdot \mathrm{grad}\,(\theta) \geq 0. \tag{2.14}$$

2.2 Thermoelasticity with Temperature-Dependent Coefficients

2.2.1 Exploitation of dissipation inequality

In the following, small deformations are considered, characterized by $\| \boldsymbol{H} \| = \sqrt{\mathrm{tr}(\boldsymbol{H}\boldsymbol{H}^{\mathsf{T}})} \ll 1$. Thus, no difference is made between the differential operators, $\mathrm{grad}\,(\cdot)$ and $\mathrm{Grad}\,(\cdot)$. Hence, the displacement gradient is given by $\boldsymbol{H} = \mathrm{grad}\,(\boldsymbol{u})$. Introducing a measure of deformation, the infinitesimal strain tensor is defined by $\boldsymbol{\varepsilon} = (\boldsymbol{H} + \boldsymbol{H}^{\mathsf{T}})/2$. Furthermore, $\boldsymbol{D}$ can be expressed in terms of the material time derivative of the infinitesimal strain tensor reading $\boldsymbol{D} = \dot{\boldsymbol{\varepsilon}}$. An additive decomposition of the infinitesimal strain $\boldsymbol{\varepsilon}$ into an elastic strain tensor $\boldsymbol{\varepsilon}_e$ and an eigenstrain tensor $\boldsymbol{\varepsilon}^*$ is assumed, i.e. $\boldsymbol{\varepsilon} = \boldsymbol{\varepsilon}_e + \boldsymbol{\varepsilon}^*$. Eigenstrains can be induced due to phase transformation, initial strains, or as in this context, by thermal expansion. In general, the free energy depends on the infinitesimal strain $\boldsymbol{\varepsilon}$, the temperature θ, the temperature gradient $\boldsymbol{g} = \mathrm{grad}\,(\theta)$, and a set $\underline{\alpha}$ of internal state variables, i.e. $\psi = \psi(\boldsymbol{\varepsilon}, \theta, \boldsymbol{g}, \underline{\alpha})$. Describing a thermoelastic material behavior, the internal state variables are not considered, and the free energy is given by

$$\psi = \psi(\boldsymbol{\varepsilon}, \theta, \boldsymbol{g}). \tag{2.15}$$

For the material time derivative of the free energy density, one obtains

$$\dot{\psi} = \frac{\partial \psi}{\partial \boldsymbol{\varepsilon}} \cdot \dot{\boldsymbol{\varepsilon}} + \frac{\partial \psi}{\partial \theta}\dot{\theta} + \frac{\partial \psi}{\partial \boldsymbol{g}} \cdot \dot{\boldsymbol{g}}. \tag{2.16}$$

Thus, the Clausius-Duhem inequality reads

$$(\boldsymbol{\sigma} - \rho\frac{\partial \psi}{\partial \boldsymbol{\varepsilon}}) \cdot \dot{\boldsymbol{\varepsilon}} - \rho(\eta + \frac{\partial \psi}{\partial \theta})\dot{\theta} - \rho\frac{\partial \psi}{\partial \boldsymbol{g}} \cdot \dot{\boldsymbol{g}} - \frac{1}{\theta}\boldsymbol{q} \cdot \boldsymbol{g} \geq 0. \tag{2.17}$$

Since the dissipation inequality is linear in $\dot{\boldsymbol{\varepsilon}}$, $\dot{\theta}$, and $\dot{\boldsymbol{g}}$, the potential relations

$$\boldsymbol{\sigma} = \rho\frac{\partial \psi}{\partial \boldsymbol{\varepsilon}}, \qquad \eta = -\frac{\partial \psi}{\partial \theta}, \tag{2.18}$$

are necessary in order to ensure $\delta \geq 0$ for the arbitrary rates $\dot{\boldsymbol{\varepsilon}}$ and $\dot{\theta}$. Furthermore, the free energy cannot depend on $\boldsymbol{g}$, i.e. $\partial\psi/\partial\boldsymbol{g} = \boldsymbol{0}$. Thus, the reduced dissipation inequality reads (Coleman and Gurtin, 1967)

$$-\frac{1}{\theta}\boldsymbol{q} \cdot \boldsymbol{g} \geq 0. \tag{2.19}$$

2.2.2 Heat conduction

Combining the balance of internal energy in Equation (2.10) with the Legendre transformation in Equation (2.13) and the potential relations in Equation (2.18) yields

$$\rho\theta\dot{\eta} + \operatorname{div}(\boldsymbol{q}) - \rho r = 0. \tag{2.20}$$

Using Equation (2.18), the material time derivative of the entropy is given by

$$\dot{\eta} = -\frac{\partial^2 \psi}{\partial \theta \partial \varepsilon} \cdot \dot{\varepsilon} - \frac{\partial^2 \psi}{\partial \theta^2} \dot{\theta} \tag{2.21}$$

leading to

$$-\rho\theta \frac{\partial^2 \psi}{\partial \theta \partial \varepsilon} \cdot \dot{\varepsilon} - \rho\theta \frac{\partial^2 \psi}{\partial \theta^2} \dot{\theta} + \operatorname{div}(\boldsymbol{q}) - \rho r = 0, \tag{2.22}$$

(Haupt, 2002). By defining the heat capacity for constant strains as

$$c_d = -\theta \frac{\partial^2 \psi}{\partial \theta^2}, \tag{2.23}$$

the differential equation for the heat conduction is derived

$$\rho c_d \dot{\theta} = \rho\theta \frac{\partial^2 \psi}{\partial \theta \partial \varepsilon} \cdot \dot{\varepsilon} - \operatorname{div}(\boldsymbol{q}) + \rho r. \tag{2.24}$$

2.2.3 Constitutive modeling

The previously discussed balance equations hold independently of the material behavior. They constitute a system of underdetermined equations, since fields are still included but not yet defined, such as the stress tensor and the internal energy. This system cannot be solved without additional constitutive relations that characterize a specific material behavior reflecting experimental observations (Müller, 1985; Truesdell and Toupin, 1960). Regarding, e.g., the previously discussed differential equation of heat conduction, a constitutive law for the heat flux $\boldsymbol{q}$ has to be chosen. Therein, Fourier's law of heat conduction is often considered, which fulfills the reduced dissipation inequality from Equation (2.19). In this context, the heat flux vector is given in terms of the heat conduction tensor $\boldsymbol{K}$ and the temperature gradient as $\boldsymbol{q} = -\boldsymbol{K}\boldsymbol{g}$

(Bertram, 2005). For the special case of a thermal isotropic material, $\boldsymbol{K} = \kappa \boldsymbol{I}$, and Fourier's law reduces to $\boldsymbol{q} = -\kappa \boldsymbol{g}$, with κ denoting the coefficient of heat conduction.

Experimental investigations of both pure and fiber-reinforced thermoset materials reveal a highly temperature-dependent but less pronounced frequency-dependent material behavior. Hence, the viscoelastic effects are neglected, and a thermoelastic material model is considered. A detailed discussion is given in Section 4.1.3. In order to capture the temperature-dependency of the material, the physical properties are modeled as functions of the temperature θ. Thus, according to Equation (2.15), the free energy is specified by

$$\rho\psi = \frac{1}{2}\boldsymbol{\varepsilon} \cdot \mathbb{C}(\theta)[\boldsymbol{\varepsilon}] - \boldsymbol{G}(\theta) \cdot \boldsymbol{\varepsilon} + h(\theta). \tag{2.25}$$

Herein, the quantity $\mathbb{C}(\theta)$ denotes the stiffness tensor, $\boldsymbol{G}(\theta)$ represents thermal stresses with respect to an unstrained state, and $h(\theta)$ represents an additional function of temperature that is related to the specific heat capacity (Schapery, 1967; 1968). Using Equation (2.18), the Cauchy stress tensor $\boldsymbol{\sigma}$ and the entropy η are given by

$$\boldsymbol{\sigma} = \rho\frac{\partial\psi}{\partial\boldsymbol{\varepsilon}} = \mathbb{C}(\theta)[\boldsymbol{\varepsilon}] - \boldsymbol{G}(\theta), \tag{2.26}$$

$$\eta = -\frac{\partial\psi}{\partial\theta} = -\frac{1}{2\rho}\boldsymbol{\varepsilon} \cdot \frac{\partial\mathbb{C}(\theta)}{\partial\theta}[\boldsymbol{\varepsilon}] + \frac{1}{\rho}\frac{\partial\boldsymbol{G}(\theta)}{\partial\theta} \cdot \boldsymbol{\varepsilon} - \frac{1}{\rho}\frac{\partial h(\theta)}{\partial\theta} \tag{2.27}$$

Considering the additive decomposition and Equation (2.26), the following relation is obtained

$$\boldsymbol{\varepsilon}^* = \boldsymbol{\varepsilon} - \boldsymbol{\varepsilon}_e = \boldsymbol{\varepsilon} - \mathbb{C}(\theta)^{-1}[\boldsymbol{\sigma}] = \mathbb{C}(\theta)^{-1}[\boldsymbol{G}(\theta)], \tag{2.28}$$

resulting in

$$\boldsymbol{G}(\theta) = \mathbb{C}(\theta)[\boldsymbol{\varepsilon}^*]. \tag{2.29}$$

An expansion of Equation (2.25) into a Taylor series with respect to a reference temperature, i.e. $\theta = \theta_0$, results in

$$\begin{aligned}\rho\psi = \frac{1}{2}\boldsymbol{\varepsilon}\cdot(\mathbb{C}(\theta_0) + \left.\frac{\partial \mathbb{C}(\theta)}{\partial \theta}\right|_{\theta_0}\Delta\theta + \mathcal{O}(\Delta\theta^2))[\boldsymbol{\varepsilon}] \\ -(\boldsymbol{G}(\theta_0) + \left.\frac{\partial \boldsymbol{G}(\theta)}{\partial \theta}\right|_{\theta_0}\Delta\theta + \left.\frac{\partial^2 \boldsymbol{G}(\theta)}{\partial \theta^2}\right|_{\theta_0}\Delta\theta^2 + \mathcal{O}(\Delta\theta^3))\cdot\boldsymbol{\varepsilon} \\ +h(\theta_0) + \left.\frac{\partial h(\theta)}{\partial \theta}\right|_{\theta_0}\Delta\theta + \left.\frac{\partial^2 h(\theta)}{\partial \theta^2}\right|_{\theta_0}\Delta\theta^2 + \mathcal{O}(\Delta\theta^3),\end{aligned} \tag{2.30}$$

with $\Delta\theta = \theta - \theta_0$. It is assumed that the stiffness $\mathbb{C}(\theta)$ depends linearly on the temperature

$$\mathbb{C}(\theta) = \mathbb{C}(\theta_0) + \left.\frac{\partial \mathbb{C}(\theta)}{\partial \theta}\right|_{\theta_0}\Delta\theta = \mathbb{C}_1 + \mathbb{C}_2\Delta\theta, \tag{2.31}$$

and the thermal stress contribution $\boldsymbol{G}(\theta)$ is a quadratic function of the temperature

$$\begin{aligned}\boldsymbol{G}(\theta) &= \boldsymbol{G}(\theta_0) + \left.\frac{\partial \boldsymbol{G}(\theta)}{\partial \theta}\right|_{\theta_0}\Delta\theta + \left.\frac{\partial^2 \boldsymbol{G}(\theta)}{\partial \theta^2}\right|_{\theta_0}\Delta\theta^2 \\ &= \boldsymbol{G}_1 + \boldsymbol{G}_2\Delta\theta + \boldsymbol{G}_3\Delta\theta^2.\end{aligned} \tag{2.32}$$

With Equation (2.29) and a Taylor series expansion with respect to $\theta = \theta_0$ for the eigenstrains, the eigenstress contribution reads

$$\boldsymbol{G}(\theta) = \mathbb{C}(\theta)[\boldsymbol{\varepsilon}^*(\theta_0) + \left.\frac{\partial \boldsymbol{\varepsilon}^*(\theta)}{\partial \theta}\right|_{\theta_0}\Delta\theta + \left.\frac{\partial^2 \boldsymbol{\varepsilon}^*(\theta)}{\partial \theta^2}\right|_{\theta_0}\Delta\theta^2]. \tag{2.33}$$

At the reference temperature θ_0, a strain-free state is assumed, i.e. $\boldsymbol{\varepsilon}^*(\theta_0) = \mathbf{0}$, and thus, $\boldsymbol{G}_1 = \mathbf{0}$. Furthermore, the first derivative of the eigenstrains with respect to θ is identified as a thermal expansion, i.e. $\partial\boldsymbol{\varepsilon}^*(\theta)/\partial\theta = \boldsymbol{\alpha}(\theta)$, with $\boldsymbol{\alpha}(\theta)$ denoting the second-order tensor of thermal expansion. Thus, Equation (2.33) yields

$$\boldsymbol{G}(\theta) = \mathbb{C}(\theta)[\boldsymbol{\alpha}(\theta_0)\Delta\theta + \left.\frac{\partial \boldsymbol{\alpha}}{\partial \theta}\right|_{\theta_0} \Delta\theta^2] = \mathbb{C}(\theta)[\boldsymbol{\alpha}_2\Delta\theta + \boldsymbol{\alpha}_3\Delta\theta^2]. \quad (2.34)$$

To be consistent with the notation in Equations (2.31) and (2.32), coefficients in front of $\Delta\theta$ are denoted by index 2, while coefficients in front of $\Delta\theta^2$ are labeled by index 3.

In the following, three special cases are in the field of interest. In this view, the additional contribution $h(\theta) = \rho h_c(\Delta\theta - \theta \ln(\theta/\theta_0))$ is used, with the constant heat capacity h_c (Bertram and Krawietz, 2012).

Linear thermoelasticity with temperature-dependent coefficients. Using the Taylor series expansion of the stiffness and the eigenstress contribution from Equations (2.31) and (2.34), respectively, the stress and entropy expressions read

$$\boldsymbol{\sigma} = (\mathbb{C}_1 + \mathbb{C}_2\Delta\theta)[\boldsymbol{\varepsilon}] - (\mathbb{C}_1 + \mathbb{C}_2\Delta\theta)[\boldsymbol{\alpha}_2\Delta\theta + \boldsymbol{\alpha}_3\Delta\theta^2] \quad (2.35)$$

$$\eta = -\frac{1}{2\rho}\boldsymbol{\varepsilon} \cdot \mathbb{C}_2[\boldsymbol{\varepsilon}] + \frac{1}{\rho}(\mathbb{C}_2[\boldsymbol{\alpha}_2\Delta\theta + \boldsymbol{\alpha}_3\Delta\theta^2] \quad (2.36)$$

$$+ (\mathbb{C}_1 + \mathbb{C}_2\Delta\theta)[\boldsymbol{\alpha}_2 + 2\boldsymbol{\alpha}_3\Delta\theta]) \cdot \boldsymbol{\varepsilon} - h_c \ln\frac{\theta}{\theta_0}. \quad (2.37)$$

In this context, the stiffness $\mathbb{C}(\theta)$ is assumed to be isotropic and is modeled in terms of the temperature-dependent Young's modulus, i.e. $\mathbb{C} = \mathbb{C}(E(\theta), \nu)$, with $E(\theta) = E_1 + E_2\Delta\theta$. The distribution of $E(\theta)$ is identified by the experimental investigations presented in Section 4.1.3. For an isotropic thermal expansion $\boldsymbol{\alpha}(\theta) = (\alpha_2\Delta\theta + \alpha_3\Delta\theta^2)\boldsymbol{I}$, the coefficients α_2 and α_3 are determined analogously by means of the experiments discussed in Section 4.2.3.

Linear thermoelasticity with constant coefficients. In view of a linear thermoelastic material behavior with constant coefficients, Equations (2.31) and (2.34) are reduced to

$$\mathbb{C}(\theta) = \mathbb{C}_1 = \mathbb{C}, \tag{2.38}$$

$$\boldsymbol{G}(\theta) = \mathbb{C}[\boldsymbol{\alpha}]\Delta\theta, \tag{2.39}$$

considering a strain-free initial state. Hence, the stress and the entropy are obtained by

$$\boldsymbol{\sigma} = \mathbb{C}[\boldsymbol{\varepsilon}] - \mathbb{C}[\boldsymbol{\alpha}]\Delta\theta, \tag{2.40}$$

$$\eta = \frac{1}{\rho}\mathbb{C}[\boldsymbol{\alpha}] \cdot \boldsymbol{\varepsilon} - h_c \ln \frac{\theta}{\theta_0}. \tag{2.41}$$

Hyperelasticity. For the last special case, a purely linear elastic material behavior is considered, i.e. an isothermal state at $\theta = \theta_0$. Thus, Equations (2.31) and (2.34) yield

$$\mathbb{C}(\theta) = \mathbb{C}, \tag{2.42}$$

$$\boldsymbol{G}(\theta) = \mathbf{0}, \tag{2.43}$$

for the stiffness and the thermal stress contribution. The resulting stress is given by the linear elastic Hooke's law, and the entropy vanishes identically, i.e.

$$\boldsymbol{\sigma} = \mathbb{C}[\boldsymbol{\varepsilon}], \tag{2.44}$$

$$\eta = 0. \tag{2.45}$$

Chapter 3

Microstructure Characterization of Long Fiber-Reinforced Polymers

3.1 Fiber-Reinforced Composites

3.1.1 Classification of fiber-reinforced composites

The knowledge of the overall or effective properties of a composite is of essential need for designing structural parts. These overall properties are highly influenced by the physical properties of the constituents and by the microstructure of the composite. In this context, the phase volume fraction represents the weakest information concerning the microstructure. More profound microstructural information is given by the orientation, size, shape, or spatial distribution of the phases. In order to account for these local properties and their impact on the effective behavior, it is essential to introduce characteristic length scales (Torquato, 2002). Considering a certain scale, the local properties, as well as the interactions between the phases of the composite, depend on the material properties and on the phase geometry. Hence, many different length scales can be considered, depending on the characteristic size of the constituent or phase considered. In the simplest case, two length scales are assigned to the composite, namely a macroscopic and a microscopic scale. For more complex structures, intermediate scales are needed, referred to as mesoscopic scales. The macroscopic scale of a sheet molding compound (SMC) composite is associated with the length scale of the sample, where the material behavior appears to be "homogeneous". On the smaller mesoscopic scale, the fibers tend to agglomerate and form bundled fiber structures. The distinct phases of the mesoscopic scale can be modeled by the bundled fibers and the matrix material. Typically, the microscopic length scale is smaller than the dimensions of the mesoscopic length scale but much larger than the length scale on the molecular level (Milton, 2002). On the microscopic level, the heterogeneous material behavior can be modeled by resolving each phase. In terms of the SMC composite, the different phases are given by the single fiber filaments and the surrounding matrix material. Regarding other types of composite materials, heterogeneities are represented by, e.g., the particles or different grains of a polycrystal. In heterogeneous media, phases can be determined that exhibit certain constant physical properties at the length scale considered, e.g., elastic moduli or thermal conductivity. These phases can be modeled as aggregates of the homogeneous phases occupying a subdomain Ω_γ of the total volume Ω.

In what follows, a fibrous, polymer-based composite is exclusively considered. Regarding this type of composite, a distinct phase is referred to as matrix phase and embeds glass or carbon fiber reinforcements. Focusing first on the matrix material, a possible classification of polymers is given with respect to their bonding structure, cf. Figure 3.1. This submicroscopic structure is only visible on the molecular length scale. Due to the present chemical processes, the bonding structure and chain morphology differ and consequently contribute to the thermomechanical behavior. Thus, a classification into thermosets, thermoplastics, and elastomers is commonly applied, based on the characteristic submicrostructure (Brinson and Brinson, 2015). Especially, thermosets are more resistant against higher temperatures and flexible with respect to design options, whereas thermoplastics offer the capability of reshaping after temperature treatment.

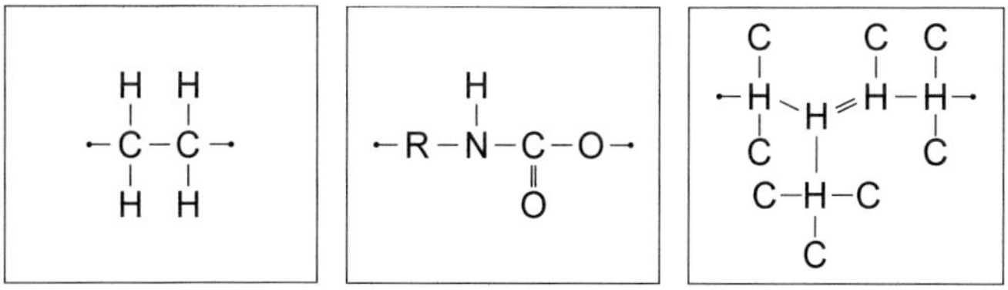

(a) (b) (c)

Figure 3.1: Characteristic polymer chain structure of a) polyethylene (thermoplastic), b) polyurethane (thermoset), and c) natural rubber (elastomer), cf. Hall (1981).

The reinforcements can be classified into continuous (Co) and discontinuous (DiCo) fibers. The former are adjusted during the production process. Thus, the fiber orientation in the final structure can be controlled in terms of the Co fibers. The adjustment of the fibers is often used to define a unidirectional fiber orientation, leading to a high anisotropy of the material properties. Hence, important design criteria concerning strength, modulus, or thermal expansion can be customized. Composites reinforced with DiCo fibers offer advantageous characteristics in the production of complex part geometries and for functional integration. Commonly, DiCo reinforcements are subdivided into long and short fibers. Figure 3.2 gives an overview of the classification of the fibrous reinforcements.

In the following, a thermoset-based composite reinforced with DiCo glass fibers is considered. The composite is produced as plaques by a sheet molding compound at the Fraunhofer Institute for Chemical Technology (ICT). The manufacturing process is described in detail by Bücheler (2018).

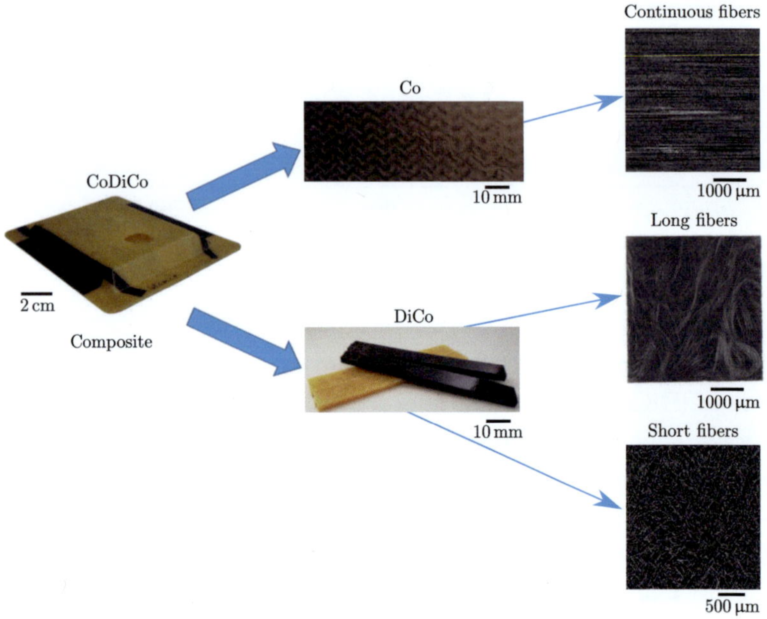

Figure 3.2: Classification of discontinuous (DiCo) and continuous (Co) fiber reinforcements. From the composite structural part (left) (demonstrator part from DFG-GRK 2078/1 project with continuous and discontinuous (CoDiCo) reinforcements), samples are extracted at different locations (middle). The samples exhibit different microstructural characteristics that can be analyzed by means of µCT scans. All µCT scans depicted are generated by Pascal Pinter of the Institute of Applied Materials (IAM-WK, KIT). Depending on the application requirements, fiber reinforcements with an endless, a short, or a long fiber length can be used (right).

3.1.2 Computed tomography scans

Computed tomography scans (µCT) are widely used to characterize the microstructure of a heterogeneous material (Kak et al., 2002). In a non-destructive method, X-rays pierce through a sample, reach a detector, and generate a radiograph. While the sample is rotated around a central axis, the procedure is performed from several angles. Herein, each radiograph depicts the absorption within the sample at a specific angle.

The absorption depends on the mass density of the different constituents. The radiographs produced are transformed into slice images and stacked as a volumetric image representing the microstructure of the considered domain of the sample. Generally, the resolution of the image can be adjusted by varying the distance between the object and the sample or the object and the detector. By implication, the resolution competes with the image size, since a higher resolution leads to a smaller scanned region and vice versa. All μCT scans within this work were performed by Pascal Pinter at the Institute of Applied Materials (IAM-WK, KIT). The collaboration took place within the context of the project DFG-GRK 2078. The scans were made on a YXLON precision computed tomography system with an open X-ray-reflection tube with a tungsten target. It uses a 2048×2048-pixel Perkin Elmer flat panel detector with a pixel pitch of $200\ \mu\mathrm{m}$. The images are acquired with an acceleration voltage of $80\ \mathrm{kV}$ and a target current of $0.1\ \mathrm{mA}$.

3.1.3 Mathematical description of a fibrous microstructure

For engineering applications, it is desirable to obtain the information on the microgeometrical configuration of a composite. This includes the fiber length distribution or, particularly, the fiber orientation distribution. For this purpose, each voxel is applied to a filtering algorithm (Pinter et al., 2018) from which fiber orientation tensors, fiber orientation distribution functions, or discrete fiber orientation histograms are determined. In this context, the resulting orientation information is limited to straight fibers. The orientation distribution for fibers with arbitrary curvatures or waviness is not clearly defined (Abdin et al., 2014). One approach is given by Komori and Marishima (1978) who approximate the orientation distribution of curved fibers by a various number of subdivided straight segments. This procedure, however, is dependent on the segment lengths and thus might lead to a computationally complex problem.

In terms of fiber orientation histograms, the fiber orientation is discretely given by the fiber orientation axis $\boldsymbol{n}_\gamma$ of a certain straight fiber γ. The fiber axis is parametrized by the angles ϑ and φ, using spherical coordinates, i.e. $\boldsymbol{n} \widehat{=} (\sin\vartheta\cos\varphi, \sin\vartheta\sin\varphi, \cos\vartheta)^\mathsf{T}$ with respect to the $\{x, y, z\}$ coordinate system, see Figure 3.3a. On the macroscopic level, homogenized quantities such as the effective stiffness $\bar{\mathbb{C}}$ are introduced. For discretely given fiber orientations $\boldsymbol{n}_\gamma$ and a parameter set $\underline{p}_\mathrm{F}$ and $\underline{p}_\mathrm{M}$ for the fiber and the matrix material, respectively, the effective stiffness can be formulated in terms of

$$\bar{\mathbb{C}} = \mathbb{f}(\underline{p}_\mathrm{F}, \underline{p}_\mathrm{M}, \boldsymbol{n}_\gamma). \tag{3.1}$$

Fiber orientation histograms represent the orientation distribution and serve as a graphical description, see Figure 3.3b.

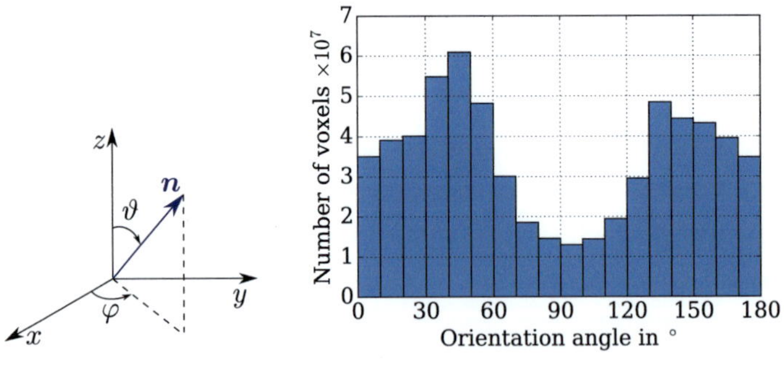

(a) **(b)**

Figure 3.3: a) Parametrized fiber orientation in spherical coordinates $\{\vartheta, \varphi\}$. b) Fiber orientation histogram of an SMC composite sample.

A more general description of the fiber orientation is given by the fiber orientation distribution function (ODF) (Advani and Tucker III, 1987). It represents the orientation density of fibers aligned in any unit direction $\boldsymbol{n}$ as a continuous distribution over the unit sphere. Using the fiber ODF $f(\boldsymbol{n})$, the effective stiffness is expressed by

$$\bar{\mathbb{C}} = \mathfrak{f}(\underline{p}_{\mathrm{F}}, \underline{p}_{\mathrm{M}}, f(\boldsymbol{n})). \tag{3.2}$$

Considering the surface on a unit sphere $S = \{\boldsymbol{n} \in \mathbb{R}^3 : \|\boldsymbol{n}\| = 1\}$, the ODF is normalized and non-negative, i.e.

$$\int_S f(\boldsymbol{n}) \,\mathrm{d}S = 1, \qquad f(\boldsymbol{n}) \geq 0, \quad \forall \boldsymbol{n} \in S, \tag{3.3}$$

with the surface element $\mathrm{d}S = \sin\vartheta \,\mathrm{d}\varphi \,\mathrm{d}\vartheta/(4\pi)$. Moreover, $f(\boldsymbol{n})$ is centrosymmtric, i.e. $f(\boldsymbol{n}) = f(-\boldsymbol{n})$ (Zheng and Zou, 2001). By means of Dirac's delta distribution function $\delta(\boldsymbol{n}, \boldsymbol{n}_\gamma)$ and N equally weighted fiber orientations $\boldsymbol{n}$, the empirical fiber ODF is defined by

$$f(\boldsymbol{n}) = \frac{1}{N} \sum_{\gamma=1}^{N} \delta(\boldsymbol{n} - \boldsymbol{n}_\gamma), \tag{3.4}$$

(Advani and Tucker III, 1987; Kanatani, 1984).

Depending on the microstructure considered, a description of the orientation state by fiber orientation distribution functions can exhibit a computational effort (Lee et al., 2003). In contrast, orientation tensors compactly represent the fiber orientation. Thus, some homogenization methods provide a direct use of orientation tensors. For this purpose, the stiffness is formulated in terms of the orientation tensor of the second- and fourth-order, $\boldsymbol{N}$ and $\mathbb{N}$, respectively,

$$\bar{\mathbb{C}} = \mathfrak{f}(\underline{p}_{\mathrm{F}}, \underline{p}_{\mathrm{M}}, \boldsymbol{N}, \mathbb{N}). \tag{3.5}$$

According to Kanatani (1984), fiber orientation tensors are moment tensors of the ODF referred to as orientation tensors of the first kind defined by

$$\boldsymbol{N} = \int_S f(\boldsymbol{n}) \boldsymbol{n} \otimes \boldsymbol{n} \, \mathrm{d}S, \qquad \mathbb{N} = \int_S f(\boldsymbol{n}) \boldsymbol{n} \otimes \boldsymbol{n} \otimes \boldsymbol{n} \otimes \boldsymbol{n} \, \mathrm{d}S. \tag{3.6}$$

In terms of an empirical fiber ODF, the orientation tensors read

$$\boldsymbol{N} = \frac{1}{N} \sum_{\gamma=1}^{N} \boldsymbol{n}_\gamma \otimes \boldsymbol{n}_\gamma, \qquad \mathbb{N} = \frac{1}{N} \sum_{\gamma=1}^{N} \boldsymbol{n}_\gamma \otimes \boldsymbol{n}_\gamma \otimes \boldsymbol{n}_\gamma \otimes \boldsymbol{n}_\gamma. \tag{3.7}$$

Herein, the fiber lengths, as well as the fiber diameters, are considered constant quantities. The orientation tensors of the first kind possess all symmetries, i.e.

$$\boldsymbol{N} = \boldsymbol{N}^{\mathsf{T}}, \qquad \mathbb{N} = \mathbb{N}^{\mathsf{T_H}} = \mathbb{N}^{\mathsf{T_R}} = \mathbb{N}^{\mathsf{T_L}} = \mathbb{N}^{\mathsf{T_M}}. \tag{3.8}$$

Furthermore, they are not linearly independent; thus, a contraction $\boldsymbol{N} = \mathbb{N}[\boldsymbol{I}]$ lowers the rank. For discussions on orientation tensors of the second and third kinds, the reader is referred to Kanatani (1984).

In the case of microstructures with fibers distributed isotropically in one plane or in space, as well as fibers aligned unidirectionally, the orientation tensors are given analytically and illustrated in Figure 3.4.

3.2 Microstructure Characterization of SMC Composites

The composite considered in this work consists of unsaturated polyester-polyurethane hybrid resin (UPPH) as the matrix material. It is reinforced with discontinuous long glass fibers of an approximately constant length of $l_{\mathrm{F}} = 25\,\mathrm{mm}$, a constant diameter of $d_{\mathrm{F}} = 12\,\mu\mathrm{m}$, and at a fiber content

$$N \mathrel{\widehat{=}} \begin{pmatrix} 1/2 & 0 & 0 \\ 0 & 1/2 & 0 \\ 0 & 0 & 0 \end{pmatrix} \qquad N \mathrel{\widehat{=}} \begin{pmatrix} 1/3 & 0 & 0 \\ 0 & 1/3 & 0 \\ 0 & 0 & 1/3 \end{pmatrix} \qquad N \mathrel{\widehat{=}} \begin{pmatrix} 1 & 0 & 0 \\ 0 & 0 & 0 \\ 0 & 0 & 0 \end{pmatrix}$$

(a) (b) (c)

Figure 3.4: Orientation tensors for a) planar isotropic, b) isotropic, and c) unidirectionally oriented fibers with illustrations of the corresponding microstructures.

of $c_{\mathrm{F}} = 22\,\mathrm{vol.\%}$. Detailed descriptions of the production process by SMC are given in Section 4.1.2. For reinforced UPPH samples, μCT scans are performed. The results of the μCT scan of a specific sample are given in Figure 3.5. The geometric dimensions of the sample are illustrated in Figure 3.5a. The orange-colored area indicates the scanned area. A reconstruction of the μCT scan as a three-dimensional volume image is shown in Figure 3.5b. Herein, the matrix material is colored in black, and the fibers are illustrated in white color. The volume element consists of $1826 \times 1887 \times 560$ voxels in the x-, y-, and z-directions and exhibits a resolution of $5.5\,\mu\mathrm{m}$. The resulting geometric dimensions of approximately $10 \times 11 \times 3\,\mathrm{mm}^3$ are illustrated in Figure 3.5b. The orientation distribution is analyzed for the μCT scan considered and depicted in Figure 3.5c. Using the HSV color map, each fiber orientation is identified by its corresponding color. The procedure for the orientation analysis is explained in detail in Pinter (2018). Fibers that are aligned purely in the x-direction are colored in red, and fibers in the y-direction in blue. Mixed colors represent mixed orientations. The color distribution indicates a slightly bigger amount of the blue color. Thus, the fibers oriented in the y-direction are slightly dominant compared to those in the x-direction. A part of the considered volume element is scanned at a higher resolution of 3 µm and depicted in Figure 3.5d. Consequently, a smaller volume region is obtained. Thus, the single fiber filaments are visible at a higher resolution at the upper surface. At the areas in between, the fibers exhibit a layered structure of fiber bundles.

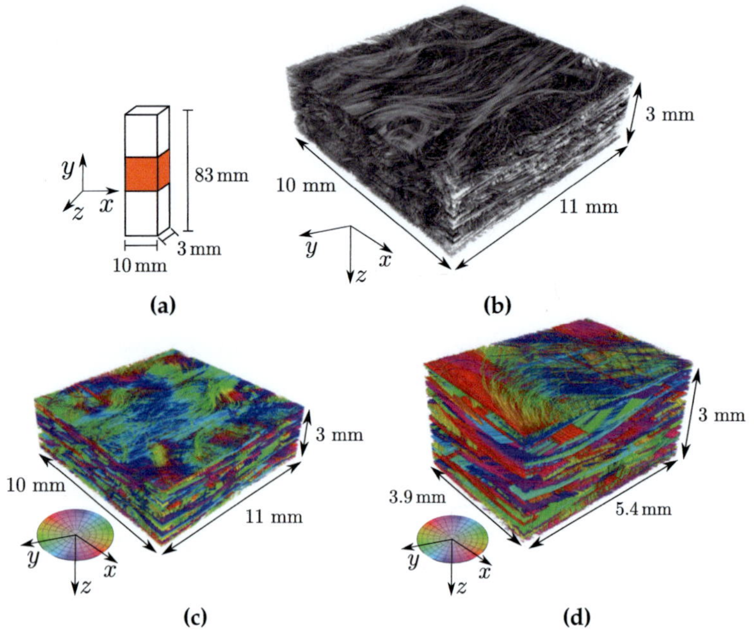

Figure 3.5: a) Geometry of the sample used for µCT analysis. b) Reconstructed µCT data for a 0° sample from a 1D plaque. c) Orientation analysis for µCT data with a resolution of 5.5 µm, and d) a resolution of 3 µm. The top layer corresponds to $z = 0$ mm, and the bottom layer to $z = 3$ mm.

Regarding both scanned volume images, slices in the x-y-plane at different positions on the z-axis are depicted in Figure 3.6. Figure 3.6a illustrates the slices for every 0.5 mm along the thickness obtained for the resolution at 5.5 µm, whereas Figure 3.6b presents the corresponding results for the higher resolution. The microstructure in the slices at the bottom and the top of the volume image, i.e. at $z \approx 0\,\mathrm{mm}$ and $z \approx 3\,\mathrm{mm}$, exhibits curved fibers. In these regions, the microstructure is governed by individual fiber filaments that have split away from the fiber bundles. In the slices in between, a strongly geometric arrangement of the fibers is detected, forming straight fiber bundles. This geometrical configuration on the microscale is typically found in SMC composites. Due to the resulting domain-like structure, SMC composites are sometimes referred to as aggregated systems (Böhm, 2004). Since these domains are characterized by roughly straight fiber segments, a mean-field method can be applied which is generally based on straight, ellipsoidal shaped single inclusions.

Figure 3.7a depicts the diagonal components of the orientation tensor of the second-order $\boldsymbol{N}$ over the thickness z for the volume image shown in Figure 3.5b. As typically studied for SMC composites, an orientation distribution with respect to the thickness is given. However, no characteristic distributions are observed that indicate a layered structure such as the core-shell distributions which are typical of LFT materials. The N_{zz} component exhibits an almost vanishing contribution and indicates a planar orientation distribution. This is reasonable since the fiber length is significantly larger compared to the thickness of the composite plaque. A comparison between the N_{xx} and N_{yy} components reveals a more dominant fiber orientation distribution with respect to the y-direction. The y-direction corresponds to the direction of the flow during the production process, cf. Section 4.1.2.

Figure 3.7b shows the distribution of the fiber volume fraction along the thickness in the z-direction. A variation along the thickness is observed. The mean value is given by $\bar{c}_{\mathrm{F}} = 0.225$ with a maximal and minimal deviation of about $\pm 20\%$.

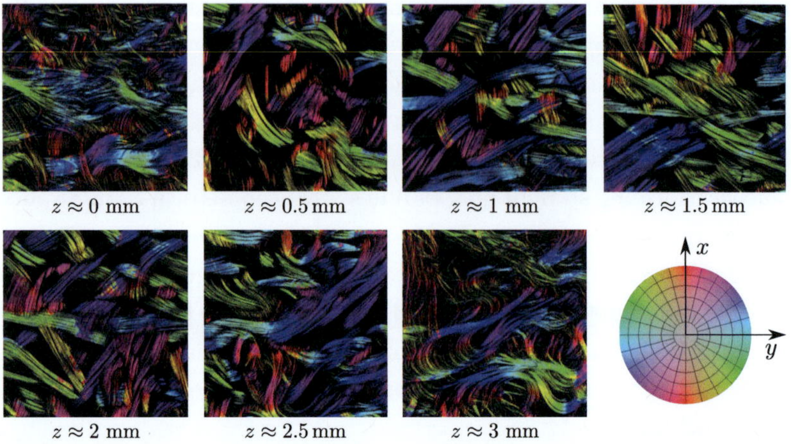

(a)

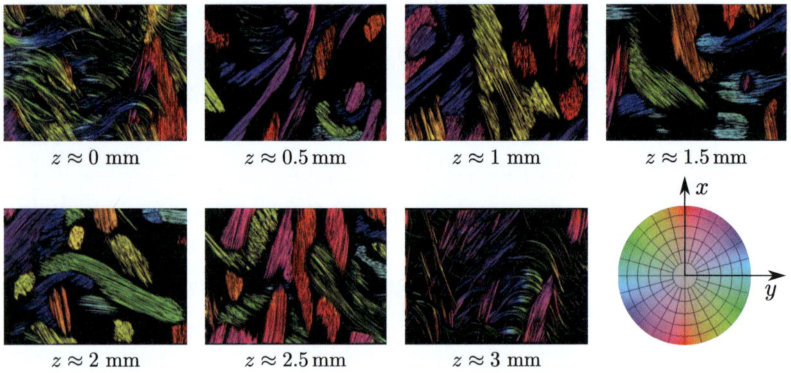

(b)

Figure 3.6: Orientation analysis for slices at different positions with respect to the z-axis in the thickness direction. Orientation analysis is performed for $0°$-oriented sample with a resolution of a) $5.5\ \mu m$ and b) $3\ \mu m$.

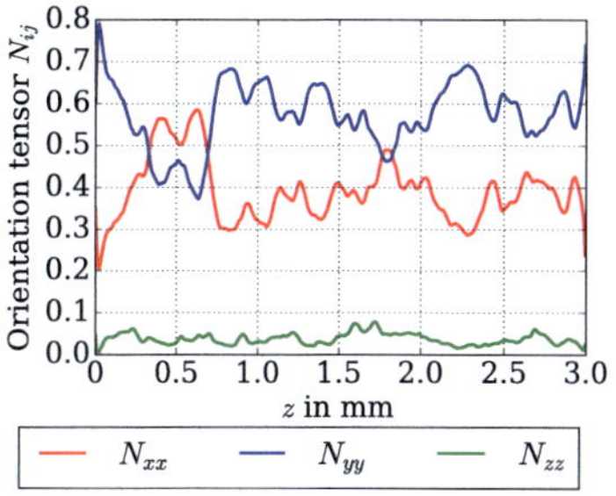

(a)

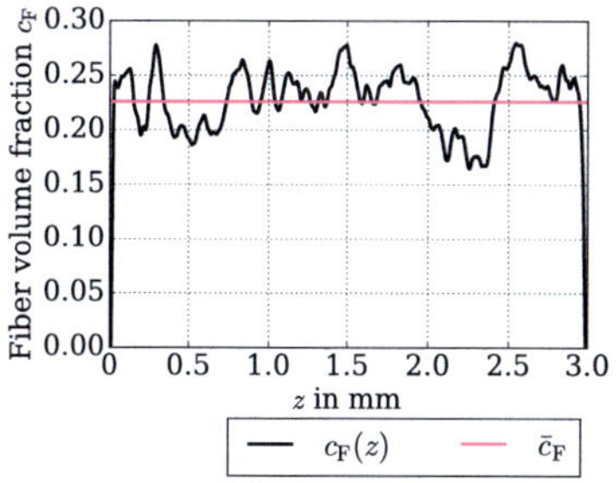

(b)

Figure 3.7: a) Diagonal components of the second-order orientation tensor $\boldsymbol{N}$ over the thickness. b) Fiber volume fraction over the thickness.

Chapter 4

Experimental Investigations of Pure and Reinforced Polymers

4.1 Dynamic Mechanical Analysis

4.1.1 Fundamentals

A dynamic mechanical analysis (DMA) is performed to investigate the thermoviscoelastic behavior of the thermoset-based composite and the corresponding matrix material (Ferry, 1980). The elastic stiffness and the viscous behavior of a sample can be determined using this measuring method. Regarding DMA, a phase-shifted material response is analyzed with respect to the dynamic excitation of the sample. First, a constant strain $\varepsilon_{\mathrm{stat}} = \varepsilon_0$ is applied to the sample, cf. Figure 4.1a. This preload is superimposed by a sinusoidal oscillation $\varepsilon_{\mathrm{dyn}} = \tilde{\varepsilon}_0 \sin(\omega t)$, with $\tilde{\varepsilon}_0$ denoting the constant amplitude, and ω the frequency of the oscillation. The total excitation consists of a static and a dynamic load reading

$$\varepsilon(t) = \varepsilon_{\mathrm{stat}} + \varepsilon_{\mathrm{dyn}} = \varepsilon_0 + \tilde{\varepsilon}_0 \sin(\omega t). \tag{4.1}$$

Hence, the material's response due to the excitation is also sinusoidal. Especially for polymers, a phase shift δ between the excitation $\varepsilon(t)$ and the response $\sigma(t)$ is observed, cf. Figure 4.1a. Thus, the resulting stress is given by

$$\sigma(t) = \sigma_0 + \tilde{\sigma}_0 \sin(\omega t + \delta) \tag{4.2}$$

(Menard, 2008). In the context of polymeric composites, the temperature-dependent behavior is of interest as well. Thus, the sample is additionally subjected to temperature load. In this context, strain-controlled tests are recommended: The increasing temperature leads to a softening of the sample. In terms of a stress-controlled excitation, the dynamic strain amplitude increases with higher temperatures, leading to increasing deformations of the sample. Thus, nonlinear, irreversible deformations of the sample are obtained at some point. However, when a constant strain load is applied, the deformation of the sample will remain in the linear, reversible region.

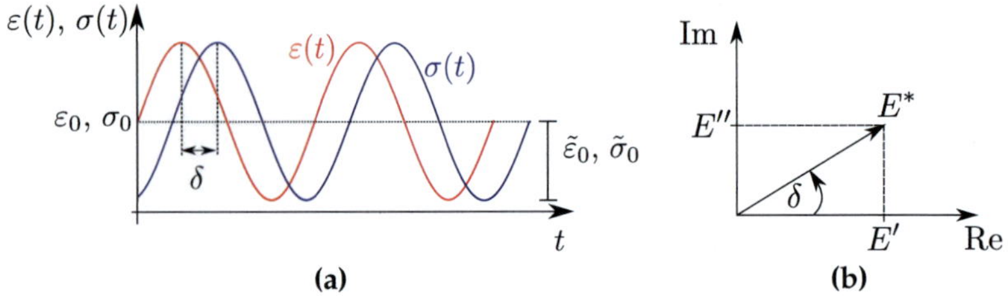

Figure 4.1: a) Sinusoidal excitation $\varepsilon(t)$ and material's response $\sigma(t)$ with amplitudes ε_0 and σ_0. The phase shift δ is given between excitation and response. b) Representation of the complex modulus E^* by the storage modulus E' as the real part and by the loss modulus E'' as the imaginary part.

The magnitude of the complex modulus E^* is determined by the ratio of the stress amplitude to the strain amplitude, reading

$$|E^*| = \frac{\tilde{\sigma}_0}{\tilde{\varepsilon}_0}. \tag{4.3}$$

As sketched in Figure 4.1b, E^* consists of a real and an imaginary part, denoted as storage modulus E' and loss modulus E'', respectively, reading

$$E^* = E' + iE''. \tag{4.4}$$

The storage modulus and the loss modulus are defined by

$$E' = |E^*| \cos\delta = \frac{\tilde{\sigma}_0}{\tilde{\varepsilon}_0} \cos\delta, \quad E'' = |E^*| \sin\delta = \frac{\tilde{\sigma}_0}{\tilde{\varepsilon}_0} \sin\delta \tag{4.5}$$

(Ehrenstein, 1999). The integral stress power consists of a stored and a dissipative part, W_{stor} and D, respectively, and is given by

$$\int \sigma(t)\dot{\varepsilon}(t)\,\mathrm{d}t = W_{\mathrm{stor}} + D \tag{4.6}$$

(Busfield and Muhr, 2003). Choosing exemplarily a quarter cycle, one obtains for the stored contribution

$$W_{\mathrm{stor}} = \tilde{\varepsilon}_0^2 E' \int_0^{\pi/(2\omega)} \sin(\omega t)\cos(\omega t)\,\mathrm{d}t = \frac{\tilde{\varepsilon}_0^2}{2} E'. \tag{4.7}$$

As Equation (4.7) reveals, W_{stor} is proportional to the storage modulus. Hence, the storage modulus represents the instantaneous elastic material response to the cyclic load. In contrast, the loss modulus characterizes the viscous behavior caused by internal friction, for instance. The loss modulus is proportional to the dissipation, reading

$$D = \tilde{\varepsilon}_0^2 E'' \int_0^{\pi/(2\omega)} \cos^2(\omega t)\,\mathrm{d}t = \frac{\pi}{4} \tilde{\varepsilon}_0^2 E''. \tag{4.8}$$

The ratio of the loss to the storage modulus is referred to as the loss factor $\tan\delta$

$$\tan\delta = \frac{\sin\delta}{\cos\delta} = \frac{E''}{E'}, \qquad \delta \in [0^\circ, 90^\circ]. \tag{4.9}$$

Thus, $E^* = E'$ and $\delta = 0^\circ$ for a purely elastic behavior. A purely viscous behavior is obtained for $E^* = E''$ and $\delta = 90^\circ$ (Ehrenstein, 1999).

Depending on the load type, the experiments can be classified into

- Temperature sweeps: varying temperature at a constant frequency
- Frequency sweeps: varying frequencies at a set temperature
- Dynamic stress-strain sweeps: increasing dynamic load amplitudes at a fixed load rate

Generally, a sweep indicates the varying part of the load applied. A time sweep, e.g., identifies the time as the varying quantity, whereas all other load parameters, such as temperature or frequency, remain constant.

Regarding temperature sweeps, the material's response is measured as a function of the temperature applied. Additionally, various thermal transitions in the polymer can be detected. In general, thermal transitions can be described by relaxation times or, equivalently, by free volume changes of polymers, including SMC materials (Dlubek et al., 2005). Based on the free volume concept, the particular size of local free volumes is related to the mobility of the molecular polymer chains. At low temperatures, the molecules are in a compressed state. Heating up the material, the free volume increases due to the bond and side chain movements, referred to as γ-transition. Ongoing heating leads to the β-transition which is characterized by movable whole side chains. By further heating, the α-transition or glass transition θ_g is finally reached. In general, the glass transition is denoted as the major transition. In Figure 4.2, an illustration of the changing bonding structures associated with the corresponding transitions is given for thermosets. For this class of materials, heating above θ_g inhibits the slipping of the chains by the cross-linked polymer chains (Brinson and Brinson, 2015). This results in the degradation and in the burning of the material. For some semi-crystalline polymers such as polypropylene, θ_g marks a lower temperature limit above which applications are recommended. Hence, θ_g is used to define the upper or lower temperature limits for possible applications. This operating range is particularly important with respect to the strength and stiffness requirements in industrial applications. In terms of purely crystalline polymers, no glass transition is observable.

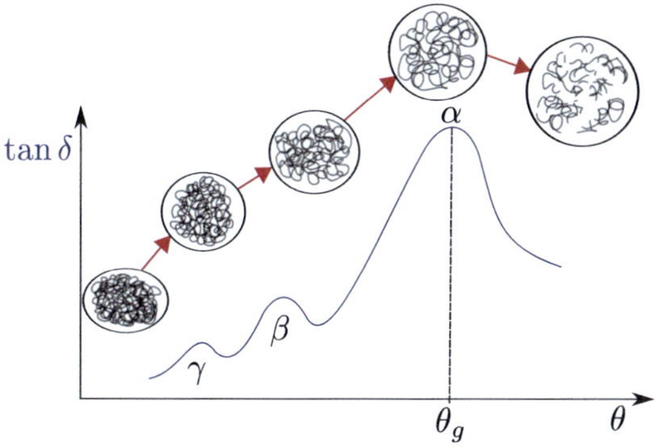

Figure 4.2: Qualitative illustration of the distribution of $\tan\delta$ over increasing temperature for a thermoset polymer. The minor transitions are denoted by γ and β, whereas the glass transition α is given at θ_g. With increasing temperature θ, the movement of polymer chains of the bonding structure increases as well. Thus, the free volume increases as illustrated by the corresponding bonding structure. After θ_g, a degradation of the chains occurs.

Frequency sweep tests focus on the viscoelastic material properties and aim at studying the effect of frequency on the temperature-driven changes of the material properties. From these tests, the relaxation behavior of polymer chains can be extracted (Lobo and Bonilla, 2003). A combination with a temperature sweep yields a so-called temperature-frequency sweep test that accounts for a varying temperature as well as a varying frequency. In this context, higher frequencies induce a more elastic-like behavior, resulting in a stiffer material behavior. Regarding the glass transition, a higher frequency shifts the observed θ_g to a higher temperature. As described in Ehrenstein (1999), there are at least five common methods to determine θ_g, e.g., the maximum loss modulus or by the 2%-method. In this work, the maximum loss factor is used to indicate the glass transition.

From dynamic stress-strain sweeps, the range of the strain amplitude that defines a load leading to a deformation within the linear elastic region can be determined. This quantifies the proper operating range of a material.

4.1.2 Experimental setup

The composite used in this thesis is produced by a sheet molding compound (Bücheler, 2018). In the first step, the pure resin material is mixed and reinforced with chopped fibers at a constant length of 2.5 cm. Afterward, the resulting semi-finished product is rolled between foils and matured for a specified time at a controlled temperature. Finally, the semi-finished product is cut into pieces and pressed into plaques. During the compression molding, temperature and pressure loads are applied to the material, leading to the curing of the resin.

In this thesis, three differently produced plaques are considered. In the first case, the semi-finished product is placed in the left third of the mold so that it flows in y-direction during the pressing process, see Figure 4.3a. Regarding the second type, the semi-finished material is positioned in the middle of the mold, leading to flow in the x- and y-directions (Figure 4.3b). With respect to the third type, the mold is completely filled with the semi-finished product so that during the production, no flow occurs, cf. Figure 4.3c. From the resulting plaques, samples were extracted at $0°$, $45°$, and $90°$ angles to the y-direction, cf. Figure 4.3d. The geometry of the samples is sketched in Figure 4.3e.

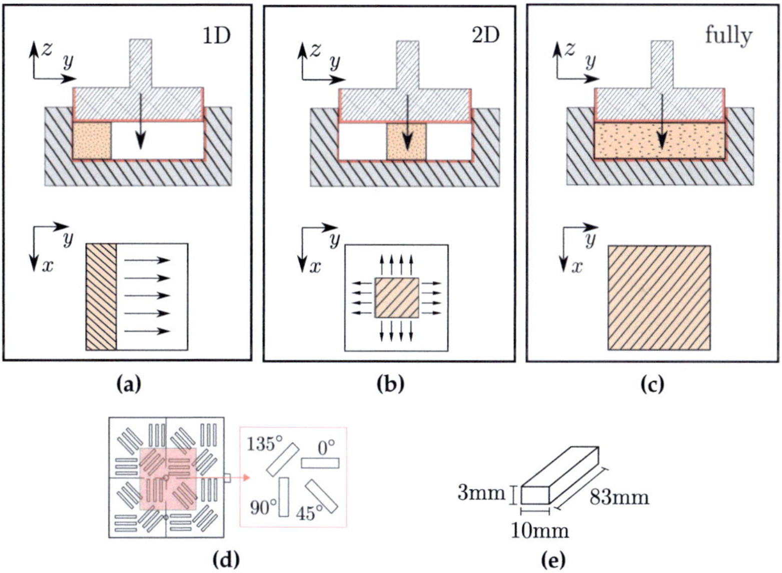

(a) (b) (c)

(d) (e)

Figure 4.3: a) Production of a composite plaque for which the semi-finished product is placed in the left third of the mold. By means of pressure and temperature load, the semi-finished product is pressed into a cured plaque. During the production process, the uncured material flows from the left part to the end of the mold, i.e. in the y-direction. b) Production of a composite plaque for which the semi-finished product is placed in the middle of the mold. During the production process, the uncured material flows in two directions, i.e. in the x- and y-directions, until it reaches the side walls of the mold. c) Production of a composite plaque for which the mold is completely filled with the semi-finished product. During the production process, no flow of the uncured material occurs. d) The cutting plan of the cured plaques. From the middle of each plaque, samples are taken at different orientations and are considered for DMA tests. e) The geometry of a DMA sample.

The DMA tests are performed by means of the testing system GABO Eplexor® 500N at the Institute of Engineering Mechanics, Chair for Continuum Mechanics, at the Karlsruhe Institute of Technology (KIT) see Figure 4.4. The tests were performed under tension mode at a controlled strain load. The temperature load is applied to the sample inside the temperature chamber. Temperature control is assured by heating elements in the temperature chamber and by instreaming liquid nitrogen. The static and the dynamic load are applied by two drivers. The static preload is applied by a displacement-controlled spindle drive, whereas the dynamic load is applied by means of a force-controlled electromagnetic shaker.

Figure 4.4: Photographs of the DMA system GABO Eplexor® 500N in the laboratory at the Institute of Engineering Mechanics, Chair for Continuum Mechanics, at the Karlsruhe Institute of Technology (KIT).

During the production process, the curing can cause stresses to be enclosed in the material. Heating up the material afterward leads to a rearrangement of the molecules, which thus relieves the trapped stresses. Furthermore, a not fully completed curing process can induce an unstable behavior in the material. Since an annealed and completely cured state of the material is desired, the samples are conditioned before the DMA tests are performed. This treatment is done by slowly heating up the samples in a tempering oven and holding the target temperature for a specified time. Depending on the sample's size and the polymer material, recommendations can be found in Ehrenstein (1999). The samples in this work are conditioned at $120°\mathrm{C}$ for $12\,\mathrm{h}$.

The following section presents the experimental results obtained for the pure and fiber-reinforced UPPH material. First, the matrix material is considered. The material behavior of an unconditioned sample is investigated by means of temperature sweep tests. Thus, conclusions can be drawn with respect to the material behavior of samples directly after the production process. Subsequently, the fiber-reinforced samples are considered for which temperature and frequency sweep tests are performed. The test parameters for the experimental procedures are listed in Table 4.1.

Static load $\varepsilon_0 = 0.1\%$	Dynamic load $\tilde{\varepsilon}_0 = 0.05\%$
Temperature $\theta \in [-30, 200]°\mathrm{C}$	Frequency $f \in [0.5, 50]\,\mathrm{Hz}$
Contact force $F_c = 5\,\mathrm{N}$	Heating/cooling rate $1\,\mathrm{K/min}$
Load cell capacity $1500\,\mathrm{N}$	

Table 4.1: Test parameters for the temperature sweep and frequency sweep tests.

4.1.3 Temperature- and frequency-dependent material behavior of pure and DiCo fiber-reinforced UPPH

Pure resin material. The production of the pure matrix samples is documented in Appendix B. The preceding tests were performed with pure resin samples. In this context, the storage modulus E' is measured at controlled room temperature of $\theta = 20°\mathrm{C}$ and at a fixed frequency load of $f = 5\,\mathrm{Hz}$. The results are illustrated in Figure 4.5. Therein, the values of E' for different samples show small deviations from each other. The mean value of all samples is $3.37\,\mathrm{GPa}$. The scatter of the measured values is given by the standard deviation from the mean value, cf. Figure 4.5. Based on the negligible scatter, the storage modulus at room temperature for the pure resin material is considered constant.

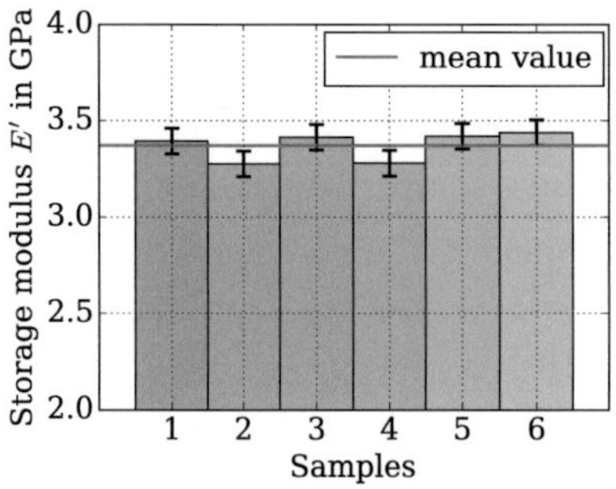

Figure 4.5: Time sweep tests for various pure UPPH samples indicate a nearly constant elastic modulus for the matrix material.

For further investigations, temperature sweep tests at constant frequency load of $f = 5\,\mathrm{Hz}$ were performed for four different samples, see Figure 4.6. In each case, two samples are from the same plaque, i.e. the description "GF12" refers to one plaque as well as "R0". Regarding all samples considered, a highly temperature-dependent behavior is observed, given by the drop in the storage modulus for increasing temperatures. Especially, for temperatures $\theta \geq 100°\mathrm{C}$, the loss factor $\tan\delta$ increases and indicates pronounced viscoelastic behavior. The peak in the $\tan\delta$ curves detects the glass transition between $155°\mathrm{C}$ and $160°\mathrm{C}$. The distribution of E' and $\tan\delta$ shows deviations between samples from different plaques, especially for $\theta \geq 50°\mathrm{C}$. This behavior indicates a sensitivity of the material properties in the production process.

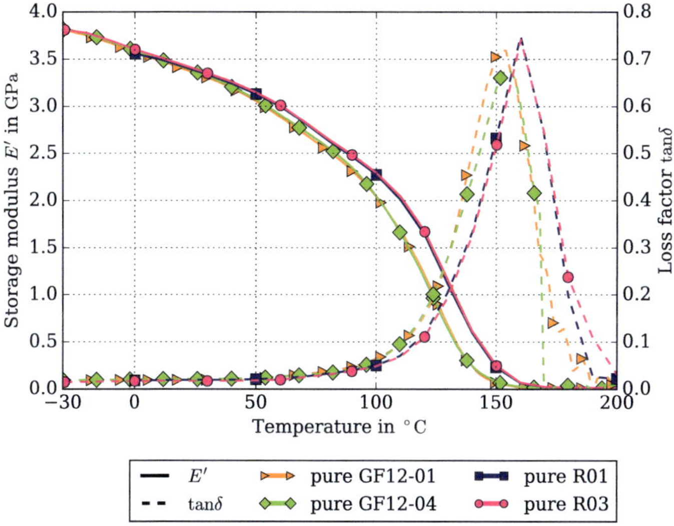

Figure 4.6: Temperature sweeps at the constant frequency $f = 5\,\mathrm{Hz}$ for selected pure UPPH samples show the temperature dependency of E' and $\tan\delta$.

For two samples of the R0-plaque, the temperature sweeps are performed at three different frequencies as depicted in Figure 4.7 on the top. A frequency-dependent behavior of E' is observed throughout the temperature region considered. The behavior is more pronounced for higher temperatures toward the glass transition. The loss factor $\tan\delta$ exhibits a fairly frequency-independent behavior for temperatures up to $100°\mathrm{C}$, illustrated in Figure 4.7 on the bottom. For higher temperatures, an increasing influence of the applied frequency on the $\tan\delta$ distribution is shown. Typically, the peak of $\tan\delta$ is shifted to a higher temperature for higher frequencies. In particular, the peak of $\tan\delta$ is given at nearly $155°\mathrm{C}$ for $f = 0.5\,\mathrm{Hz}$ whereas it is shifted by almost $18\,\mathrm{K}$ to $173°\mathrm{C}$ for $f = 50\,\mathrm{Hz}$.

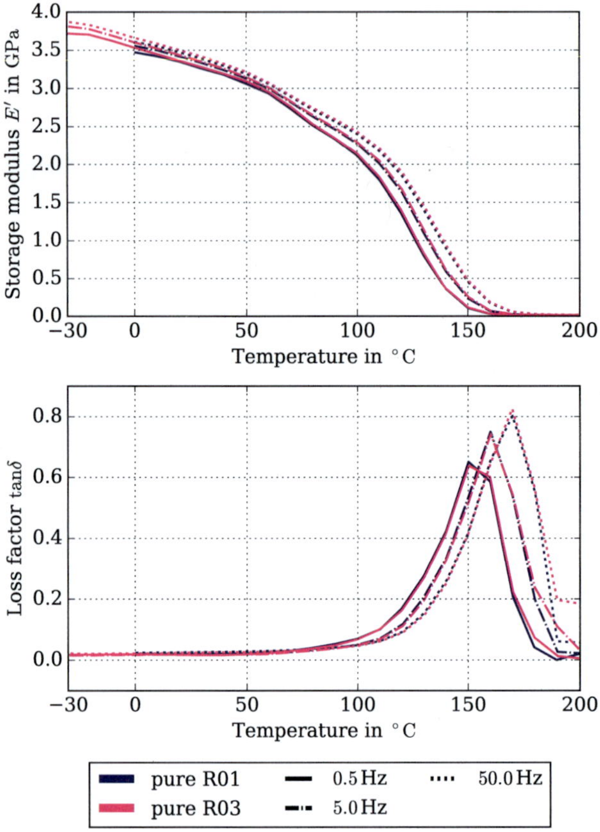

Figure 4.7: Frequency sweeps for selected pure UPPH samples. The storage modulus E' and loss factor $\tan\delta$ are depicted over temperature θ for three frequency decades.

Reinforced material. As stated previously, a conditioning of the samples before the tests are performed leads to a decreasing scatter of the experimental data. Consequently, the measured behavior of conditioned samples is not affected by the post-curing effects of the material, for instance. However, the quantitative difference between the unconditioned and conditioned material behavior is of interest. Thus, in order to investigate the behavior of the fiber-reinforced material directly after the production, temperature sweep tests have been performed using an unconditioned sample. For these investigations, a 90°-oriented sample from the 1D plaque is used. During the DMA test, the sample is first cooled down from room temperature to -30°C and kept constant at this temperature for $30\,\mathrm{min}$. Subsequently, the chamber is heated to 200°C and exposed to this temperature for $90\,\mathrm{min}$. This cyclic temperature load is repeated as sketched in Figure 4.8. Exposing the sample to a high temperature during the first cycle constitutes a conditioning of the sample. Thus, the behavior of the sample for the second and further cycles corresponds to a conditioned sample, and can be compared to the first cycle which represents the material at unconditioned state.

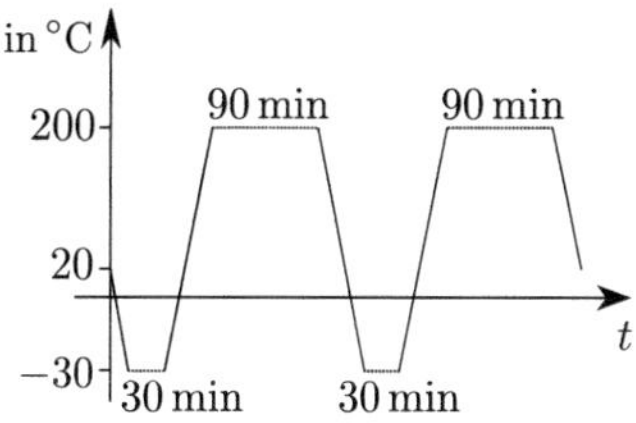

Figure 4.8: Heating and cooling cycles for a 90°-oriented, fiber-reinforced UPPH sample.

The results of the temperature sweep tests are depicted in Figure 4.9. The storage modulus over the temperature is depicted on the left y-axis, whereas the loss factor over temperature is shown with respect to the right y-axis. Regarding the two heating cycles, E' and $\tan\delta$ show negligible deviations. The same is observed for the two cooling cycles. Slight deviations are observed between the heating and the cooling phases. The test was performed for other samples, showing comparable results. Thus, no pronounced curing processes are observed directly after the production of the samples. However, in order to ensure that no process-induced behavior affects the interpretation of the results, all the samples used for further investigations have been conditioned ahead of time.

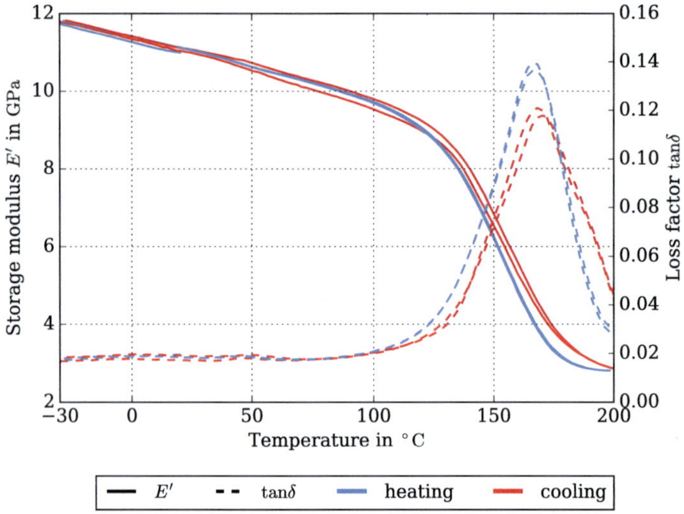

Figure 4.9: Storage modulus E' and loss factor $\tan\delta$ over temperature θ for the different heating and cooling cycles.

For the three types of differently produced plaques, the temperature sweep tests are performed for three differently oriented samples at $f = 5\,\mathrm{Hz}$. The storage and the loss modulus, as well as the loss factor over temperature, are depicted in Figure 4.10. All fiber-reinforced UPPH samples exhibit a highly temperature-dependent behavior. Especially, the storage modulus E' for all samples exhibits a drop between the maximal and minimal value of 75% to 80%. Regarding the 1D plaque, the storage modulus decreases with an increasing inclination angle between the flow direction and the orientation of the sample in the cured plaque. This indicates a fairly anisotropic behavior. In contrast, the samples of the 2D and the fully filled plaque show no distinct anisotropic behavior. The distribution of E' is approximately the same for all orientations of these samples. This indicates a planar isotropic material behavior. The loss modulus E'' of all samples is characterized by a nearly constant distribution for temperatures lower than $100°\mathrm{C}$. For temperatures above $100°\mathrm{C}$, E'' increases nonlinearly for all samples until it reaches a maximum at approximately $155°\mathrm{C}$. Quantitative deviations are observed in terms of the different plaques. For samples of the 1D plaque, E'' depends on the orientation of a sample. This is far less pronounced with increasing planar isotropic oriented samples of the 2D and fully filled plaque. The loss factor $\tan\delta$ in all samples exhibits a similar behavior as that for the loss modulus. For temperatures below $100°\mathrm{C}$, small values of nearly 0.02 are measured, whereas a strong increase is observed at higher temperatures. An increase of $\tan\delta$ indicates a distinct viscoelastic behavior, whereas the small values below $100°\mathrm{C}$ show a pronounced elastic behavior with negligible viscous effects. Compared to the results for the pure material, the glass transition temperature for the fiber-reinforced material is shifted to higher temperatures. In particular, the glass transition temperature θ_g is detected between $160°\mathrm{C}$ and $170°\mathrm{C}$, depending on the sample considered. Consequently, the operating temperature range of the reinforced thermoset material is also shifted toward higher temperatures.

In addition to the temperature sweep tests, varying the frequency allows investigating the frequency-dependent and thus viscoelastic behavior. For these investigations, three samples of the 1D plaque are considered. The frequency is varied between $0.5\,\mathrm{Hz}$ and $50\,\mathrm{Hz}$, which covers three frequency decades. Analogous to the foregoing studies, a temperature range of $-30°\mathrm{C}$ to $200°\mathrm{C}$ is considered. The results are shown in Figure 4.11. The storage modulus and the loss factor for seven frequencies are depicted over temperature. Regarding the results for E', a remarkably small influence of the different frequency loads on the measured behavior is observed for temperatures less than $100°\mathrm{C}$.

A pronounced frequency-dependent behavior is detected for temperatures higher than $100°\mathrm{C}$. Especially near θ_g, an increasing frequency yields an increased storage modulus of approximately 33% when comparing the highest and lowest frequency load for all considered samples. The frequency load affects the $\tan\delta$ distribution particularly for temperatures above $100°\mathrm{C}$. Moreover, the peak of $\tan\delta$ is shifted toward higher temperatures for increasing frequencies. In particular, the peak for the lowest frequency is shifted by nearly $20\,\mathrm{K}$ compared to the peak for the highest frequency.

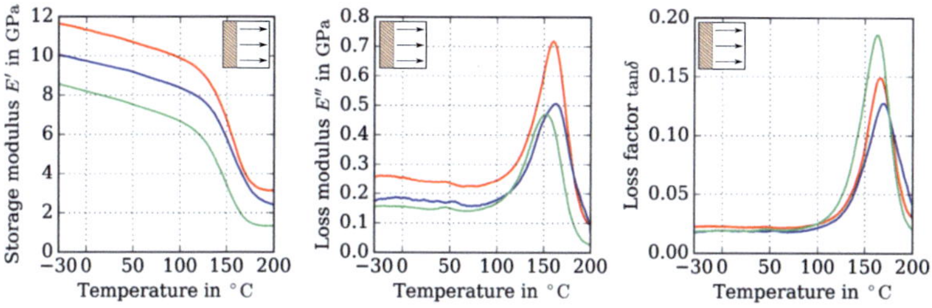

(a) 1D plaque

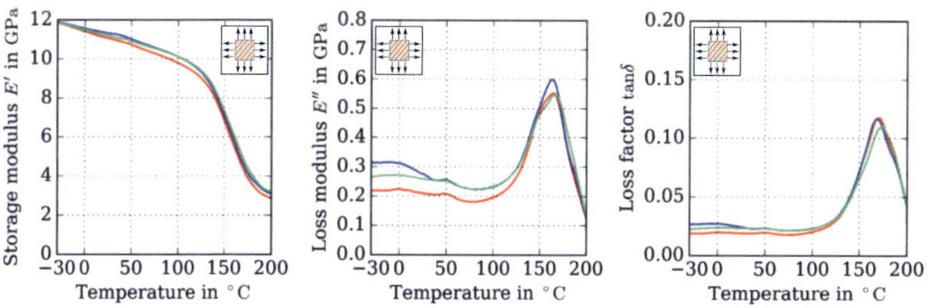

(b) 2D plaque

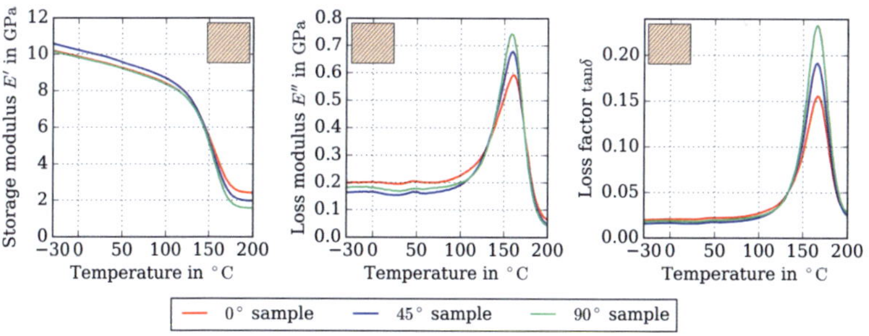

(c) Fully filled plaque

Figure 4.10: Temperature sweep tests for samples of the three differently produced plaques. Storage modulus E', loss modulus E'', and loss factor $\tan\delta$ over temperature for $0°$-, $45°$-, and $90°$-oriented samples of the a) 1D plaque, b) 2D plaque, and the c) fully filled plaque.

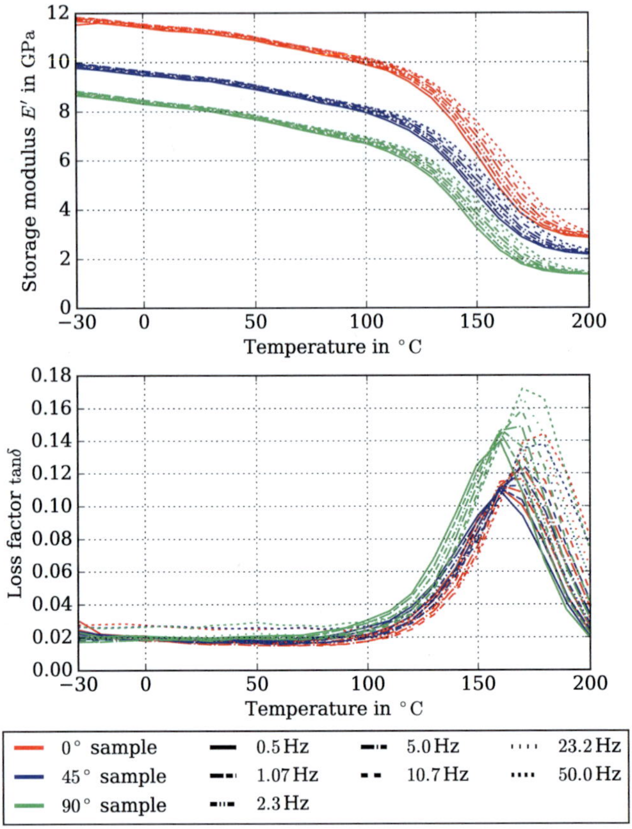

Figure 4.11: Frequency sweep tests for fiber-reinforced UPPH samples of the 1D plaque. Storage modulus E' and loss factor $\tan\delta$ of the $0°$-, $45°$-, and $90°$-oriented sample are depicted over temperature for seven frequencies in three decades.

4.1.4 Summary of results

The DMA tests performed for the pure and fiber-reinforced UPPH samples reveal a highly temperature-dependent behavior. For the entire temperature range, the E' of the composite samples is shifted by an approximately constant factor of 2.5 (90° sample) to 3.0 (0° sample) toward higher values compared to E' of the pure resin material. This shows the reinforcing effect of the fibers within the composite material. Furthermore, the pure UPPH samples show a more pronounced frequency dependency for the entire temperature region compared to the reinforced samples. Samples from differently produced plaques reveal an anisotropic behavior when the flow of the uncured resin occurs during the production process. The viscoelastic behavior is analyzed using the loss factor. For all composite samples, the loss factor exhibits a constant distribution at a small value regarding temperatures below 100°C. Increasing viscoelastic behavior is observed only for the temperatures above 100°C. These results are confirmed by the frequency sweeps since a pronounced frequency dependency is measured for the higher temperatures near glass transition temperature and above. As discussed previously, θ_g identifies the temperature operating range of a polymeric composite. The composite cannot be used above θ_g due to degradation. A typical range of application in automotive engineering does not exceed 100°C. In the range from −30°C to 100°C, viscous effects can be neglected with respect to the computation of the effective properties. Thus, the mean- and full-field methods considered in this thesis account for the thermoelastic material behavior.

4.2 Measurement of Thermal Expansion

4.2.1 Experimental setup

The DMA results in Section 4.1.3 reveal a significant temperature-dependent behavior for both the pure and fiber-reinforced UPPH material. To characterize the thermal material behavior, tests are performed to determine the coefficient of thermal expansion (CTE). These tests were carried out at the Fraunhofer Project Center (FPC) in London, Canada, in collaboration with Prof. Wood from the University of Western Ontario (Canada). The MTS Environmental Chamber 651 is used to control temperature load applied to the samples. The thermal strain is measured using the video extensometer system Imetrum IM-Lens-MT010 in combination with the MTS TestSuite™ TWE software, see Figure 4.12a. For this purpose, a speckle pattern is sprayed onto the samples. Inside the environmental chamber, the sample is fixed at the bottom end, whereas the upper end is unattached. The video extensometer camera, placed in front of the environmental chamber, recordes the pattern on the sample. Using the MTS TestSuite™ TWE software, two target points are defined on the sample. During the test, the longitudinal strain between the two target points is measured. After its calibration at room temperature, the environmental chamber is manually heated up. Measurements are performed at target temperatures of $\theta = [50, 100, 150, 200]\,^{\circ}\mathrm{C}$. To ensure a moderate heating rate of approximately $2\,\mathrm{K/min}$, the environmental chamber is heated up in discrete temperature increments of $10\,\mathrm{K}$ and held for $2\,\mathrm{min}$ at each increment, see Figure 4.12b.

At each target temperature, the temperature load is held constant for $15\,\mathrm{min}$ to ensure a homogeneous temperature distribution within the sample. Subsequently, the thermal strain is measured by recording 20 measuring points. The mean value and the corresponding standard deviations are determined based on the values measured.

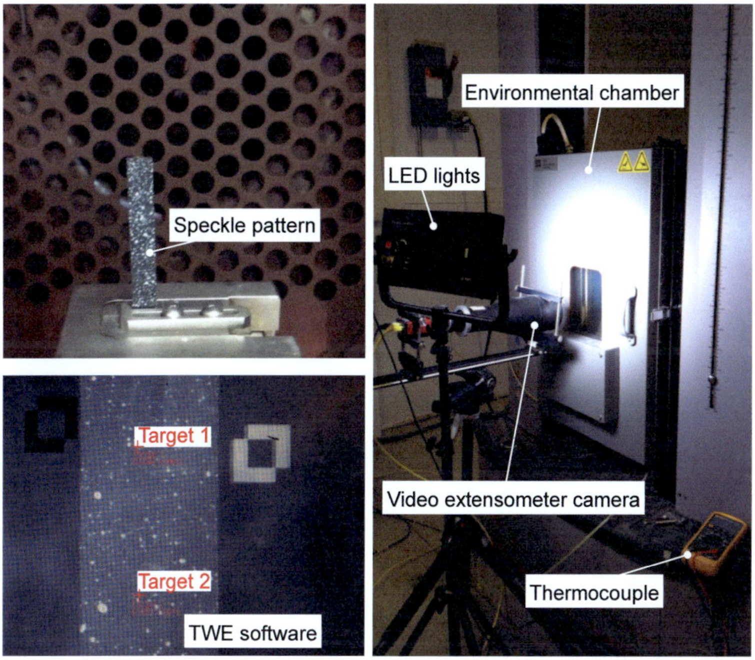

(a)

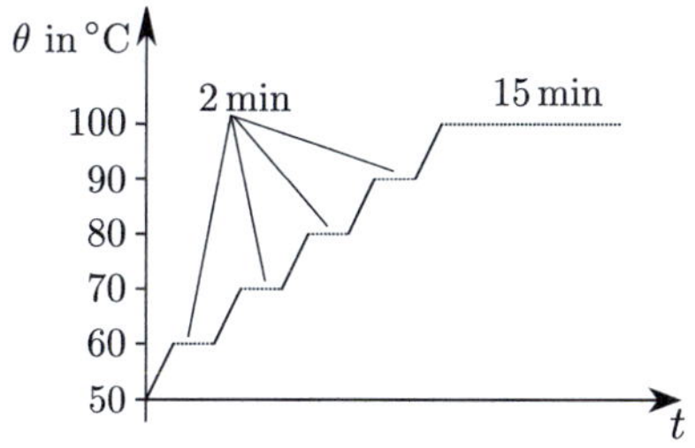

(b)

Figure 4.12: a) Experimental setup for CTE measurements. The sample is fixed at the bottom end and placed into the environmental chamber. When heating up the thermal chamber, the sample elongates. This temperature-induced expansion is tracked by a video extensometer camera. The thermal strains are measured using the MTS TestSuite™ TWE software. b) Temperature ramps during the CTE measurements, exemplarily depicted between two recorded temperature points, 50°C and 100°C.

4.2.2 Validation based on aluminum

Prior to the thermal expansion measurements of the UPPH material, tests were performed using an aluminum sample. Since the CTE of aluminum is well known from the literature, these tests serve as validation of the experimental procedure.

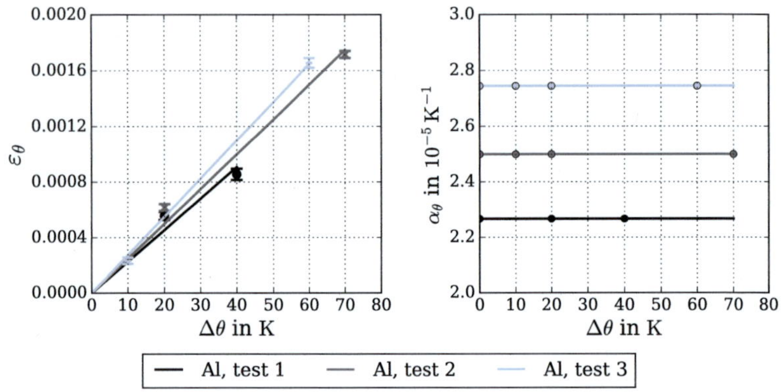

Figure 4.13: Validation tests for the CTE measurements using an aluminum sample. The experimental data of the thermal strains are captured by a linear fit (left). The corresponding CTE is given on the right. A reference temperature of about 30°C was used for the experiments.

In Figure 4.13 (left), the thermal strain ε_θ is depicted over $\Delta\theta$ for three test runs. In the temperature range of 20°C to 100°C, the distribution of the thermal strain can be considered linear (Touloukian and Ho, 1970). Thus, the least square fit of a linear function serves as an approximation of the measured thermal strain as depicted in Figure 4.13. The CTE is calculated by $\alpha_\theta = \mathrm{d}\varepsilon_\theta / \mathrm{d}\theta$. Therefore, the corresponding CTE is a constant value as shown in Figure 4.13 on the right. Herein, the determined CTE values are in the range between $2.26 \times 10^{-5}\,\mathrm{K}^{-1}$ and $2.7 \times 10^{-5}\,\mathrm{K}^{-1}$. According to data given in, e.g., Nix and MacNair (1941);

Touloukian and Ho (1970); Cverna (2002), the CTE of aluminum amounts to $2.1 - 2.51 \times 10^{-5}\,\mathrm{K}^{-1}$. Thus, the values determined for the CTE are in good agreement with values given in the literature, indicating a correct calibration of the experimental setup.

4.2.3 Thermal expansion and CTE of pure and DiCo fiber-reinforced UPPH

Pure resin material. For the pure UPPH material, two tests with two samples are performed as depicted in Figure 4.14. The upper diagrams show the measured thermal expansions ε_θ, depicted as dots. Illustrated by solid lines, a linear fit of the measuring points (left), as well as a quadratic fit (right), is shown for each sample. Since the scatter of the measured values is small, the standard deviations almost vanish. The thermal strains measured for both samples are close together. In the first test, both linear and quadratic least square fits show a good approximation of the measured data. In the case of the second UPPH sample, no experimental data at high temperatures are available. Thus, the quadratic fit is not appropriate for the second sample and is only presented for completeness. The diagrams below show the corresponding distribution of the thermal expansion coefficient α_θ. From the linear fit of the thermal strain data, a constant CTE distribution is obtained. Thus, a quadratic fit of ε_θ results in a linear distribution of α_θ. Herein, the first test shows a monotonically decreasing curve for α_θ, whereas the second test exhibits an increasing linear curve. This contradictory behavior is due to the quadratic fit of the second test, which, however, is not representative as previously mentioned. Consequently, in the following chapters, a constant CTE of the matrix material is used for computing the effective CTE of the composite material.

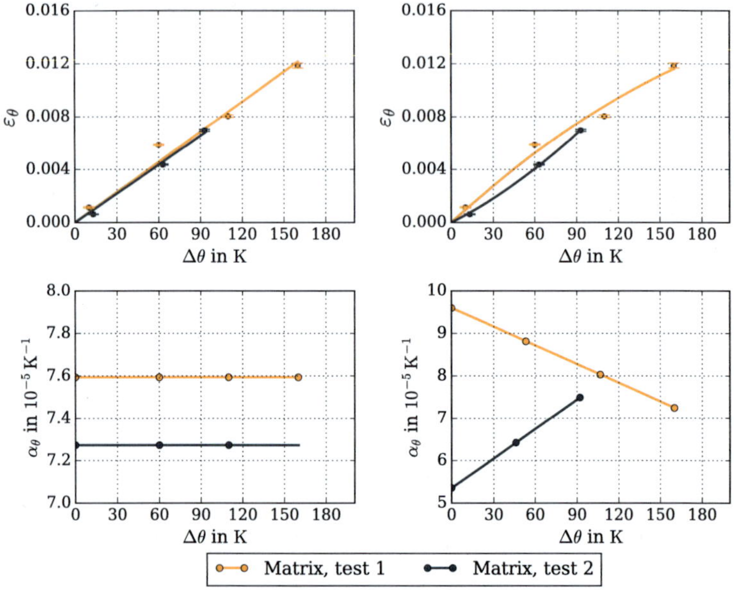

Figure 4.14: Thermal expansion measurements and resultant CTE of the pure UPPH material. The linear (left) and quadratic (right) fits to the experimentally obtained thermal strains are shown in the upper row. The corresponding constant (left) and linear (right) fits for the resultant CTE is shown in the lower row. A reference temperature of about $\theta_0 = 35°\mathrm{C}$ was used for the experiments. The quadratic fit of the second test is not appropriate and only presented for completeness.

Reinforced material. Of the 1D plaque, three differently oriented samples are considered. The experimental data represented by dots and their corresponding standard deviation are depicted in the upper diagrams of Figure 4.15. In the left diagram, a linear fit to the experimental data is given. The right diagram shows a quadratic fit for one dataset chosen of each orientation of a sample.

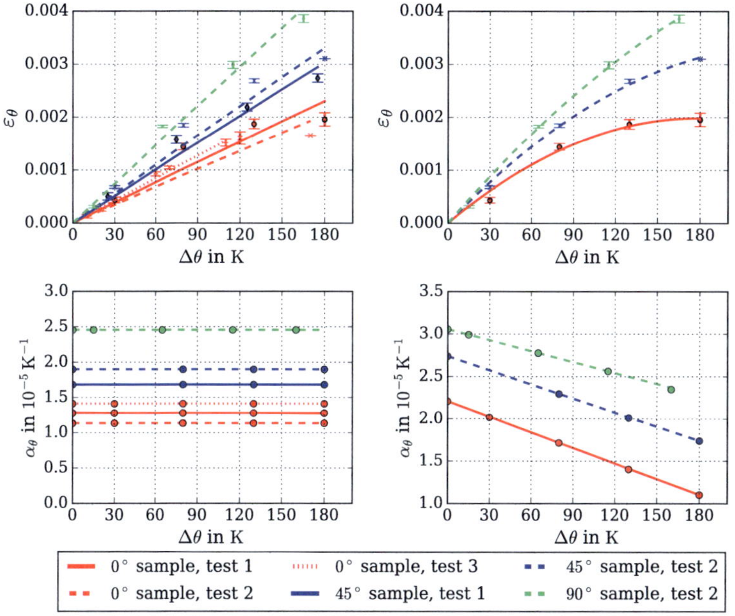

Figure 4.15: Thermal expansion measurements and resultant CTE of the 0°-, 45°-, and 90°-oriented, fiber-reinforced UPPH samples from the 1D plaque. Linear (left) and quadratic (right) fits to the experimentally obtained thermal strains are shown for each sample in the upper row. The corresponding constant (left) and linear (right) fits for the CTE is given in the diagrams below. A reference temperature of about $\theta_0 = 25°C$ was used in the experiments.

The linear fit for each orientation and sample exhibits the deviations of the data points measured. In contrast, the quadratic fits are in good agreement with the experimentally obtained thermal expansions. This subsequently motivates the modeling of the CTE as a function of the temperature.

In general, an anisotropic behavior for the thermal strain is observed. The sample with a $0°$ orientation exhibits the least thermal expansion, whereas the $90°$-oriented sample shows the highest values. The diagrams below illustrate the corresponding coefficients of thermal expansion. The CTE values of the fiber-reinforced material are 3 to 5 times smaller than those of the pure UPPH material, indicating the influence of the glass fibers which exhibit a CTE value of $\alpha_\theta = 0.5 \times 10^{-5}\,\mathrm{K}^{-1}$ (Granta Design, 2016). As for the linear fit of ε_θ, the CTE values scatter between nearly $1.1 \times 10^{-5}\,\mathrm{K}^{-1}$ and $2.5 \times 10^{-5}\,\mathrm{K}^{-1}$ for the differently oriented samples. The monotonically decreasing linear functions for α_θ are obtained from the quadratic fit of ε_θ.

4.2.4 Summary of results

In this thesis, a rather new material class is considered. Thus, the experimental data needed for modeling the material behavior are hardly available. As discussed previously, the UPPH composite exhibits a pronounced temperature-dependent behavior. In this view, a thermoelastic material model needs to be considered. Therefore, tests are performed to measure the thermal expansion of pure and fiber-reinforced UPPH samples. The CTE is then determined from these tests. In terms of the pure resin material, a fairly linear behavior of the thermal strains with increasing temperature is measured. Thus, a constant CTE is obtained for the matrix material. In contrast, the fiber-reinforced UPPH samples exhibit a pronounced nonlinear distribution of ε_θ with increasing temperatures. Hence, a linear temperature-dependent function is obtained for the CTE of the fiber-reinforced samples. Moreover, the CTE of the composite material exhibits anisotropic behavior.

Chapter 5

Homogenization for Fiber-Reinforced Polymers

5.1 Mean-Field Homogenization

5.1.1 Fundamental equations

As analyzed in the previous chapters, the SMC composite exhibits a highly heterogeneous microstructure. This significantly affects the material behavior on the macroscopic scale. The transition from the microscopic to the macroscopic scale is accomplished using homogenization techniques (Dvorak, 2012). The aim of the homogenization techniques is to determine an equivalent medium on a macroscopic level that exhibits comparable material properties to those of the heterogeneous medium. In the classical homogenization theories, a representative volume element (RVE) is introduced. An RVE is considered as the minimal material volume that contains enough information on the microstructure to statistically represent the composite material. On the surface of the RVE, macroscopically homogeneous boundary conditions are applied, leading to bounds or estimates of the overall stiffness or compliance. To obtain results that are independent of a specific choice of boundary conditions and of the absolute size of the constituents of the composite, an RVE is required to be sufficiently large (Hill, 1963). Simultaneously, it is required that the RVE is sufficiently small compared to the macroscopic dimensions. Consequently, the appropriate size of an RVE is closely related to the type of composite and the underlying microstructure. Thus, determining it can be a challenging task. In the special case of a composite containing spherical inclusions, for instance, Drugan and Willis (1996) determined that the RVE size is sufficiently large by choosing twice the reinforcement diameter. In some cases, an RVE cannot be found. This is due to the great variety of the microstructural characteristics from one realization to another. In cases such as these, use is made of statistical or stochastic volume elements (SVE) (Ostoja-Starzewski, 2006).

Section 3.2 introduces an SMC composite with discontinuous glass fibers. A sample of the composite material represents a realization w of a stochastic process. The collection of all possible realizations is referred to as an ensemble. The phases of uniform material properties are identified within a specific realization or sample of volume Ω. In terms of an N-phase composite, each phase γ occupies a subvolume Ω_γ, with $\gamma = 1, \ldots, N$. The indicator function $\mathcal{I}_\gamma(\boldsymbol{x}, w)$ is introduced by

$$\mathcal{I}_\gamma(\boldsymbol{x}, w) = \begin{cases} 1, & \boldsymbol{x} \in \Omega_\gamma, \\ 0, & \boldsymbol{x} \notin \Omega_\gamma, \end{cases} \tag{5.1}$$

describing the geometry of the microstructure (Nemat-Nasser and Hori, 1993). Furthermore, the heterogeneous microstructure is characterized by a function $\Psi(\boldsymbol{x}, w)$ that describes certain material properties. The distribution of the material properties within an N-phase composite with phase-wise constant material properties Ψ_γ is obtained by

$$\Psi(\boldsymbol{x}, w) = \sum_{\gamma=1}^{N} \mathcal{I}_\gamma(\boldsymbol{x}, w) \Psi_\gamma(w). \tag{5.2}$$

In the context of the ergodic hypothesis, averaging the material properties over all realizations of the ensemble is equivalent to the average results, considering the volume of one realization w_0 in the infinite-volume limit. Thus, the ensemble average of the function $\Psi(\boldsymbol{x})$ is identified by the volume average in the limit of an infinitely large volume

$$\langle\Psi(\boldsymbol{x})\rangle = \lim_{N\to\infty}\frac{1}{N}\sum_{w=1}^{N}\Psi(\boldsymbol{x},w) = \lim_{\Omega\to\infty}\frac{1}{\Omega}\int_{\Omega}\Psi(\boldsymbol{x},w_0)\,\mathrm{d}\Omega. \tag{5.3}$$

Subsequently, a specific realization is considered; thus, w_0 is omitted in the following equations. The volume fraction is related to the indicator function by

$$\langle\mathcal{I}_\gamma(\boldsymbol{x})\rangle = c_\gamma, \tag{5.4}$$

with $\sum_{\gamma=1}^{N} c_\gamma = 1$. From Equations (5.2) and (5.3), in terms of piecewise constant material properties, one obtains

$$\langle\Psi(\boldsymbol{x})\rangle = \langle\sum_{\gamma=1}^{N}\mathcal{I}_\gamma\Psi_\gamma\rangle = \sum_{\gamma=1}^{N}\Psi_\gamma\langle\mathcal{I}_\gamma(\boldsymbol{x})\rangle = \sum_{\gamma=1}^{N}\Psi_\gamma c_\gamma. \tag{5.5}$$

To ensure a consistent transition between microscopic and macroscopic quantities, the equivalence between the macroscopic power and the volume average of the microscopic power is presumed (Hill, 1963). Considering the special case of linear elasticity, the same holds true for the macroscopic strain energy and the volume average of the microscopic strain energy. These conditions, known as the Hill-Mandel conditions, read

$$\langle\boldsymbol{\sigma}\cdot\dot{\boldsymbol{\varepsilon}}\rangle = \langle\boldsymbol{\sigma}\rangle\cdot\langle\dot{\boldsymbol{\varepsilon}}\rangle, \tag{5.6}$$

$$\langle\boldsymbol{\sigma}\cdot\boldsymbol{\varepsilon}\rangle = \langle\boldsymbol{\sigma}\rangle\cdot\langle\boldsymbol{\varepsilon}\rangle, \tag{5.7}$$

and are valid independently of the constitutive assumptions. The stress and strain fields are decomposed into a mean value and a fluctuating contribution, i.e. $\boldsymbol{\sigma}(\boldsymbol{x}) = \bar{\boldsymbol{\sigma}} + \tilde{\boldsymbol{\sigma}}(\boldsymbol{x})$ and $\varepsilon(\boldsymbol{x}) = \bar{\varepsilon} + \tilde{\varepsilon}(\boldsymbol{x})$. The Hill-Mandel condition implies that the averaged stress and strain fluctuations do not contribute to the internal work, i.e.

$$\langle \tilde{\boldsymbol{\sigma}} \cdot \tilde{\varepsilon} \rangle = 0. \tag{5.8}$$

In addition, the effective stress and strain fields on a macroscopic scale, $\bar{\boldsymbol{\sigma}}$ and $\bar{\varepsilon}$, are defined as the volume average of the local stress and strain field, respectively,

$$\bar{\boldsymbol{\sigma}} := \langle \boldsymbol{\sigma}(\boldsymbol{x}) \rangle = \frac{1}{\Omega} \int_{\Omega} \boldsymbol{\sigma}(\boldsymbol{x}) \, \mathrm{d}\Omega = \frac{1}{\Omega} \sum_{\gamma=1}^{N} \int_{\Omega_\gamma} \boldsymbol{\sigma}(\boldsymbol{x}) \, \mathrm{d}\Omega = \sum_{\gamma=1}^{N} c_\gamma \boldsymbol{\sigma}_\gamma, \tag{5.9}$$

$$\bar{\varepsilon} := \langle \varepsilon(\boldsymbol{x}) \rangle = \frac{1}{\Omega} \int_{\Omega} \varepsilon(\boldsymbol{x}) \, \mathrm{d}\Omega = \frac{1}{\Omega} \sum_{\gamma=1}^{N} \int_{\Omega_\gamma} \varepsilon(\boldsymbol{x}) \, \mathrm{d}\Omega = \sum_{\gamma=1}^{N} c_\gamma \varepsilon_\gamma. \tag{5.10}$$

In this context, fibers are regarded as heterogeneities on the microstructure; thus, voids, cracks, or pores are not considered.

In the following, the effective thermoelastic properties of an SMC composite are derived. By means of the constitutive equation

$$\boldsymbol{\sigma} = \mathbb{C}[\varepsilon] - \boldsymbol{G}, \tag{5.11}$$

cf. Equation (2.26), the effective stress on the macroscopic scale is obtained based on the mechanical loads and eigenstress contributions by

$$\bar{\boldsymbol{\sigma}} = \langle \mathbb{C}[\varepsilon] \rangle - \langle \boldsymbol{G} \rangle. \tag{5.12}$$

The local strains within the microstructure are commonly related to the globally applied external strains and eigenstrain contribution. To this end, localization tensors of the second- and fourth-order are introduced

$$\varepsilon = \mathbb{A}[\bar{\varepsilon}] - \boldsymbol{a}, \tag{5.13}$$

with $\mathbb{A}$ denoting the mechanical strain localization tensor of the fourth-order and $\boldsymbol{a}$ representing the eigenstrain localization tensor of the second-order. The local strains evoked by a uniform overall strain load are determined by $\mathbb{A}$, which is, moreover, independent of the applied eigenstrains. Furthermore, $\boldsymbol{a}$ corresponds to the strains within an RVE subjected to a uniform eigenstrain load, e.g., a uniform temperature change $\Delta\theta$ while no deformation is applied, i.e. under strain-free boundary conditions. Generally, the mechanical strain localization tensor exhibits the minor symmetries but not necessarily the major symmetry, i.e. $\mathbb{A} = \mathbb{A}^{\mathsf{T_L}} = \mathbb{A}^{\mathsf{T_R}} \neq \mathbb{A}^{\mathsf{T_H}}$. The eigenstrain localization tensor is symmetric as well, i.e. $\boldsymbol{a} = \boldsymbol{a}^{\mathsf{T}}$. Averaging Equation (5.13), the normalization conditions are directly determined

$$\langle \mathbb{A} \rangle = \mathbb{I}^{\mathsf{S}}, \quad \langle \boldsymbol{a} \rangle = \mathbf{0}. \tag{5.14}$$

Combining Equations (5.12) and (5.13) yields

$$\bar{\boldsymbol{\sigma}} = \langle \mathbb{C}\mathbb{A} \rangle [\bar{\boldsymbol{\varepsilon}}] - \langle \mathbb{C}[\boldsymbol{a}] + \boldsymbol{G} \rangle. \tag{5.15}$$

From the latter equation, the effective stiffness is determined by

$$\bar{\mathbb{C}} = \langle \mathbb{C}\mathbb{A} \rangle. \tag{5.16}$$

Considering the Mandel-Levin relation (Levin, 1976; Mandel, 1965), the effective eigenstress contribution $\bar{\boldsymbol{G}} = \langle \mathbb{C}[\boldsymbol{a}] + \boldsymbol{G} \rangle$ can be expressed in terms of the mechanical localization tensor, reading

$$\bar{\boldsymbol{G}} = \langle \mathbb{A}^{\mathsf{T_H}}[\boldsymbol{G}] \rangle, \tag{5.17}$$

cf. Appendix A.2.2. Using Equation (2.28), the effective eigenstrain tensor $\bar{\boldsymbol{\varepsilon}}^*$ is determined by

$$\bar{\boldsymbol{\varepsilon}}^* = \bar{\mathbb{C}}^{-1} \langle \mathbb{A}^{\mathsf{T_H}} \mathbb{C}[\boldsymbol{\varepsilon}^*] \rangle. \tag{5.18}$$

Thermally induced eigenstrains caused by a uniform temperature load are referred to as thermal strains, $\varepsilon^* = \varepsilon_\theta$. Relating the thermal eigenstrains to the thermal expansion coefficient as given in Section 2.2.3, the effective thermal expansion is calculated by

$$\bar{\boldsymbol{\alpha}} = \bar{\mathbb{C}}^{-1} \langle \mathbb{A}^{\mathsf{T_H}} \mathbb{C}[\boldsymbol{\alpha}] \rangle. \tag{5.19}$$

Equivalently, using the stress localization tensor $\mathbb{B}$, the effective thermal expansion can be determined by $\bar{\boldsymbol{\alpha}} = \langle \mathbb{B}^{\mathsf{T_H}}[\boldsymbol{\alpha}] \rangle$.

Assumptions with respect to $\mathbb{A}$ lead to various estimates for determining the effective material properties. In the context of fibrous composites, the fibers can be modeled as ellipsoidal-shaped inclusions. The geometry of the fibers is accounted for by means of Hill's polarization tensor $\mathbb{P}_0$ for ellipsoidal inclusions. For special choices of the ellipsoid's half-axes, the special cases of spherical, needle-, and disk-shaped inclusions can be derived. In these cases, Hill's polarization tensor is analytically given, cf. Willis (1977). Taking the long discontinuous fibers of the SMC composite into account, the fibers are modeled as needle-shaped inclusions using the notation by Walpole (1969). The general form of $\mathbb{P}_0$ for an ellipsoid, as well as the analytical expressions for the special case of spherical and needle-shaped inclusions, are given in Appendix A.1 (Castañeda and Suquet, 1998). Introducing a reference medium with an isotropic reference stiffness $\mathbb{C}_0$, the strain localization tensor for solving the boundary value problem is given in the normalized form by

$$\mathbb{A} = \mathbb{M} \langle \mathbb{M} \rangle^{-1}, \qquad \mathbb{M} = (\mathbb{P}_0^{-1} + \delta\mathbb{C})^{-1}, \tag{5.20}$$

(Willis, 1981; Walpole, 1969), with $\delta\mathbb{C} = \mathbb{C} - \mathbb{C}_0$. The eigenstrain localization tensor is related to the strain localization tensor by

$$\boldsymbol{a} = \mathbb{A} \langle \mathbb{M}[\boldsymbol{G}] \rangle - \mathbb{M}[\boldsymbol{G}], \tag{5.21}$$

cf. Appendix A.2.3.

The SMC material considered in this work is regarded as a two-phase composite with isotropic phases. For a phase γ, the elastic stiffness $\mathbb{C}_\gamma$ is formulated in terms of the first and second fourth-order projectors, $\mathbb{P}_1 = \boldsymbol{I} \otimes \boldsymbol{I}/3$ and $\mathbb{P}_2 = \mathbb{I}^{\mathsf{S}} - \mathbb{P}_1$, respectively,

$$\mathbb{C}_\gamma = \frac{E_\gamma}{1 - 2\nu_\gamma}\mathbb{P}_1 + \frac{E_\gamma}{2(1 + \nu_\gamma)}\mathbb{P}_2, \tag{5.22}$$

with the engineering constants E_γ denoting Young's modulus, and ν_γ denoting Poisson's ratio of the corresponding phase. In terms of temperature loads, the isotropic tensor of thermal expansion for the CTE α_γ of a specific phase reads

$$\boldsymbol{\alpha}_\gamma = \alpha_\gamma \boldsymbol{I}. \tag{5.23}$$

Within this thesis, two cases are distinguished. On the one hand, the full- and mean-field homogenization methods are compared regarding a fixed temperature, Section 6.2.1 and Section 6.3.2. Thus, the engineering constants and the CTE values are considered temperature-independent. On the other hand, the temperature dependency of the stiffness and of the CTE is considered. This is motivated by the experimental results presented in Section 4.1.3. In order to capture the temperature dependency, the stiffness of the matrix material is modeled by means of the temperature-dependent Young's modulus, i.e. $\mathbb{C}_{\mathrm{M}} = \mathbb{C}(E_{\mathrm{M}}(\theta), \nu)$ with $E_{\mathrm{M}}(\theta) = E_{\mathrm{M}1} + E_{\mathrm{M}2}\Delta\theta$ and $\Delta\theta = \theta - \theta_0$. Due to the lack of experimental data, Poisson's ratio is assumed to be temperature-independent. Analogously, the CTE of the matrix material is modeled in terms of temperature-dependent coefficients by $\alpha_{\mathrm{M}} = \alpha_{\mathrm{M}1} + \alpha_{\mathrm{M}2}\Delta\theta$ with respect to a reference temperature θ_0.

5.1.2 Mean-field homogenization based on a variable reference stiffness

Based on the Hashin-Shtrikman method, a two-step homogenization method with variable reference stiffness is presented. In view of this, the formulations from Walpole (1969) and Willis (1981) are used, accounting for eigenstrains.

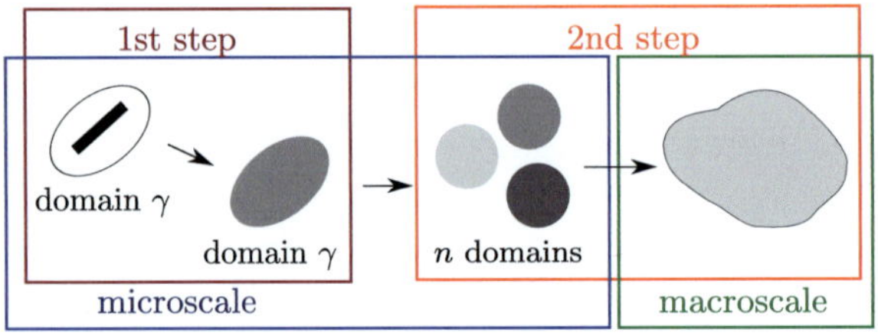

Figure 5.1: Schematic illustration of the Hashin-Shtrikman based two-step homogenization.

A two-phase composite material is assumed to consist of N ellipsoidal fibers embedded in a polymeric matrix. In a first step, the microstructure is partitioned into domains with fibers of a specific orientation surrounded by the matrix material, cf. Figure 5.1. For these domains with unidirectional fibers, the effective elastic and thermal properties are determined by applying the Hashin-Shtrikman (HS) variational principle and computing the upper and lower HS bounds. The upper HS bounds of the first step are derived by choosing the fiber stiffness as reference stiffness, whereas the lower HS bounds are computed in terms of the matrix stiffness as reference stiffness. A detailed derivation is given in Appendix A.2.1. The effective elastic material properties can then be calculated for each domain γ by

$$\bar{\mathbb{C}}_{\gamma}^{\mathrm{UD}+} = \mathbb{C}_{\gamma} + c_{\mathrm{M}}\delta\mathbb{C}_{\gamma}(\mathbb{I}^{\mathrm{S}} + c_{\gamma}\mathbb{P}_{\gamma}^{\mathrm{UD}}\delta\mathbb{C}_{\gamma})^{-1}, \tag{5.24}$$

$$\bar{\mathbb{C}}_{\gamma}^{\mathrm{UD}-} = \mathbb{C}_{\mathrm{M}} - c_{\gamma}\delta\mathbb{C}_{\gamma}(\mathbb{I}^{\mathrm{S}} - c_{\mathrm{M}}\mathbb{P}_{\gamma}^{\mathrm{UD}}\delta\mathbb{C}_{\gamma})^{-1}, \tag{5.25}$$

and the effective thermal expansion is determined by

$$\begin{aligned}\bar{\boldsymbol{\alpha}}_\gamma^{\mathrm{UD}+} &= (\bar{\mathbb{C}}_\gamma^{\mathrm{UD}+})^{-1}(\mathbb{C}_\gamma[\boldsymbol{\alpha}_\gamma] + c_{\mathrm{M}}(\mathbb{I}^{\mathsf{S}} + c_\gamma \mathbb{P}_\gamma^{\mathrm{UD}} \delta\mathbb{C}_\gamma)^{-\mathrm{T_H}})[\delta\boldsymbol{\beta}_\gamma], \\ \bar{\boldsymbol{\alpha}}_\gamma^{\mathrm{UD}-} &= (\bar{\mathbb{C}}_\gamma^{\mathrm{UD}-})^{-1}(\mathbb{C}_{\mathrm{M}}[\boldsymbol{\alpha}_{\mathrm{M}}] + c_\gamma(\mathbb{I}^{\mathsf{S}} - c_{\mathrm{M}} \mathbb{P}_\gamma^{\mathrm{UD}} \delta\mathbb{C}_\gamma)^{-\mathrm{T_H}})[-\delta\boldsymbol{\beta}_\gamma], \end{aligned} \quad (5.26)$$

with $\delta\mathbb{C}_\gamma = \mathbb{C}_{\mathrm{M}} - \mathbb{C}_\gamma$ and $\delta\boldsymbol{\beta}_\gamma = \mathbb{C}_{\mathrm{M}}[\boldsymbol{\alpha}_{\mathrm{M}}] - \mathbb{C}_\gamma[\boldsymbol{\alpha}_\gamma]$. The matrix material is denoted by a subscripted M, whereas the index $\gamma = 1, \dots, N$ denotes the modeled fibers. The fiber volume fraction is given by c_γ, for which $\sum_{\gamma=1}^{N} c_\gamma + c_{\mathrm{M}} = 1$ holds. Hill's polarization tensor for the special case of unidirectional fibers $\mathbb{P}^{\mathrm{UD}}$ is analytically given in Equation (A.7) (Willis, 1986).

In the second step, the total effective stiffness is computed by averaging over all domains

$$\begin{aligned}\bar{\mathbb{C}}^{\mathrm{HS}\pm} &= \langle \frac{1}{c_{\mathrm{F}}} \bar{\mathbb{C}}^{\mathrm{UD}\pm} \mathbb{A} \rangle = \mathbb{C}_0 - \mathbb{P}_0^{-1} + \langle \frac{1}{c_{\mathrm{F}}} (\mathbb{P}_0^{-1} + \bar{\mathbb{C}}^{\mathrm{UD}\pm} - \mathbb{C}_0)^{-1} \rangle^{-1} \\ &= \mathbb{C}_0 - \mathbb{P}_0^{-1} + \langle \mathbb{A}^* \rangle_{\mathrm{oa}}^{-1}, \end{aligned} \quad (5.27)$$

with the strain localization $\mathbb{A}$ given in Equation (5.20). The effective thermal expansion is approximated by

$$\begin{aligned}\bar{\boldsymbol{\alpha}}^{\mathrm{HS}\pm} &= (\bar{\mathbb{C}}^{\mathrm{HS}\pm})^{-1} \langle \frac{1}{c_{\mathrm{F}}} \mathbb{A}^{\mathsf{T_H}} \bar{\mathbb{C}}^{\mathrm{UD}\pm} [\bar{\boldsymbol{\alpha}}^{\mathrm{UD}\pm}] \rangle \\ &= (\bar{\mathbb{C}}^{\mathrm{HS}\pm})^{-1} \langle \mathbb{M} \rangle^{-\mathsf{T_H}} \langle \frac{1}{c_{\mathrm{F}}} \mathbb{M}^{\mathsf{T_H}} [\bar{\boldsymbol{\beta}}^{\mathrm{UD}\pm}] \rangle \\ &= (\bar{\mathbb{C}}^{\mathrm{HS}\pm})^{-1} \langle \mathbb{M} \rangle_{\mathrm{oa}}^{-\mathsf{T_H}} \langle \boldsymbol{\beta}^* \rangle_{\mathrm{oa}}. \end{aligned} \quad (5.28)$$

A detailed derivation is given in Appendix A.2.1. For the modeling of the material behavior in the foregoing equations, for simplicity, an isotropic two-point correlation of the domains is assumed. Thus, $\mathbb{P}_0$ denotes the spherical polarization tensor, which is given in Equation (A.5). The reference stiffness $\mathbb{C}_0$ is formulated in terms of a variable parameter k by

$$\mathbb{C}_0 = (1-k)\mathbb{C}_\mathrm{M} + k\,\mathbb{C}_\mathrm{F}, \qquad k \in [0,1]. \tag{5.29}$$

Since the mechanical properties of a composite are bounded by those of its constituents, $k \in [0,1]$ is a physically motivated choice, leading to $\mathbb{C}_0 = \mathbb{C}_\mathrm{M}$ for $k=0$, and $\mathbb{C}_0 = \mathbb{C}_\mathrm{F}$ for $k=1$. Thus, the reference stiffness takes values in the range of the matrix and the fiber stiffness.

In Equation (5.27), $\mathbb{P}_0$ and $\mathbb{C}_0$ exhibit the isotropic material symmetry whereas $\bar{\mathbb{C}}^{\mathrm{UD}\pm}$ is a transversely isotropic tensor. Thus, the expression within the brackets, i.e. $\mathbb{A}^*$, is of the transverse isotropic structure as well, with the symmetry axis in $\boldsymbol{e}_1$-direction. According to Advani and Tucker III (1987), the orientation average $\langle\mathbb{A}^*\rangle_\mathrm{oa}$ and $\langle\boldsymbol{\beta}^*\rangle_\mathrm{oa}$ can be evaluated using the orientation tensors of second- and fourth-order, $\boldsymbol{N}$ and $\mathbb{N}$,

$$\begin{aligned}\langle\mathbb{A}^*\rangle_\mathrm{oa} &= b_1\mathbb{N} + b_2(\boldsymbol{N}\otimes\boldsymbol{I} + \boldsymbol{I}\otimes\boldsymbol{N}) + b_3(\boldsymbol{N}\square\boldsymbol{I} + (\boldsymbol{N}\square\boldsymbol{I})^{T_R} \\ &\qquad + (\boldsymbol{I}\square\boldsymbol{N})^{T_H} + (\boldsymbol{I}\square\boldsymbol{N})^{T_R}) + b_4\boldsymbol{I}\otimes\boldsymbol{I} + b_5\mathbb{I}^\mathrm{S},\end{aligned} \tag{5.30}$$

$$\langle\boldsymbol{\beta}^*\rangle_\mathrm{oa} = \beta_1\boldsymbol{N} + \beta_2\boldsymbol{I}, \tag{5.31}$$

with the coefficients

$$\begin{aligned} b_1 &= A^*_{1111} + A^*_{2222} - 2A^*_{1122} - 4A^*_{1212}, \\ b_2 &= A^*_{1122} - A^*_{2233}, \\ b_3 &= A^*_{1212} + (A^*_{2233} - A^*_{2222})/2, \\ b_4 &= A^*_{2233}, \\ b_5 &= A^*_{2222} - A^*_{2233}, \end{aligned} \tag{5.32}$$

and

$$\begin{aligned} \beta_1 &= \lambda_{\max}(\boldsymbol{\beta}^*) - \lambda_{\min}(\boldsymbol{\beta}^*), \\ \beta_2 &= \lambda_{\min}(\boldsymbol{\beta}^*), \end{aligned} \tag{5.33}$$

where $\lambda_{\min}$ and $\lambda_{\max}$ denote the minimal and maximal eigenvalue of $\boldsymbol{\beta}^*$ (Camacho et al., 1990; Mlekusch, 1999).

5.2 Full-Field Homogenization Based on Fast Fourier Transformation

In order to determine the effective as well as local properties of composite materials with a complex microstructure, Moulinec and Suquet (1994; 1998) developed an iterative method based on fast Fourier transformation (FFT). Since the overall behavior of a composite is largely influenced by its microstructure, a detailed resolution of the constituents on the microscale plays an important role in predicting the effective properties. Based on computed X-ray tomography, the microstructure of a sample is provided by images containing pixels to which mechanical properties can be assigned. By stacking these images, a three-dimensional image is derived that represents a volumetric microstructure. A periodic, statistic, or unit cell containing a heterogeneous material with position-dependent stiffness $\mathbb{C}(\boldsymbol{x})$ is loaded by a prescribed strain $\bar{\boldsymbol{\varepsilon}}$. The periodic boundary value problem for unknown displacement fluctuations $\tilde{\boldsymbol{u}}$ reads

$$\begin{aligned} \operatorname{div}(\boldsymbol{\sigma}(\boldsymbol{x})) &= \mathbf{0} && \forall \boldsymbol{x} \in \Omega, \\ \boldsymbol{\sigma}(\boldsymbol{x}) &= \mathbb{C}(\boldsymbol{x})[\bar{\boldsymbol{\varepsilon}} + \operatorname{sym}(\operatorname{grad}(\tilde{\boldsymbol{u}}(\boldsymbol{x})))] && \forall \boldsymbol{x} \in \Omega, \\ \tilde{\boldsymbol{u}}(\boldsymbol{x}) &\quad \text{periodic} && \forall \boldsymbol{x} \in \partial\Omega, \\ \boldsymbol{\sigma}(\boldsymbol{x})\boldsymbol{n}(\boldsymbol{x}) &\quad \text{anti-periodic} && \forall \boldsymbol{x} \in \partial\Omega. \end{aligned} \tag{5.34}$$

As an auxiliary problem for a volume element with homogeneous stiffness $\mathbb{C}_0$, the stress polarization $\boldsymbol{\tau}(\boldsymbol{x})$ is introduced by

$$\boldsymbol{\tau}(\boldsymbol{x}) = (\mathbb{C}(\boldsymbol{x}) - \mathbb{C}_0)[\boldsymbol{\varepsilon}(\boldsymbol{x})]. \tag{5.35}$$

The boundary value problem can be reduced to an integral equation for $\boldsymbol{\varepsilon}(\boldsymbol{x})$, known as the Lippmann-Schwinger equation (LSE) (Kröner, 1977). The periodic LSE in real and Fourier space, respectively, is given by

$$\boldsymbol{\varepsilon}(\boldsymbol{x}) = \bar{\boldsymbol{\varepsilon}} - \mathbb{G}_0(\boldsymbol{x}) * \boldsymbol{\tau}(\boldsymbol{x}), \tag{5.36}$$

$$\hat{\boldsymbol{\varepsilon}}(\boldsymbol{\xi}) = -\hat{\mathbb{G}}_0(\boldsymbol{\xi})\hat{\boldsymbol{\tau}}(\boldsymbol{\xi}), \qquad \forall \boldsymbol{\xi} \neq \mathbf{0},\, \hat{\boldsymbol{\xi}}(\mathbf{0}) = \bar{\boldsymbol{\varepsilon}}, \tag{5.37}$$

with the convolution $\mathbb{G}_0(\boldsymbol{x}) * \boldsymbol{\tau}(\boldsymbol{x}) = \int_\Omega \mathbb{G}_0(\boldsymbol{x} - \boldsymbol{y})[\boldsymbol{\tau}(\boldsymbol{y})]\,\mathrm{d}\Omega$. In the latter equation, the periodic Green operator $\mathbb{G}_0$ is associated with the reference material $\mathbb{C}_0$, cf. Kabel et al. (2014). A µCT image is discretized into $N_1 \times N_2$ pixels in the two-dimensional case or into $N_1 \times N_2 \times N_3$ voxels for three-dimensional problems. Thus, the continuous form of the Fourier transform given in Equation (5.37) is replaced by the discrete Fourier transform (DFT) using trigonometric interpolation. According to Moulinec and Suquet (1998), the algorithm for the DFT reads

Initialization:		$\boldsymbol{\varepsilon}^0(\boldsymbol{x}_d) = \bar{\boldsymbol{\varepsilon}}, \quad \forall \boldsymbol{x}_d \in \Omega$
Iterate $i+1$:	i)	$\boldsymbol{\tau}^i(\boldsymbol{x}_d) = (\mathbb{C}(\boldsymbol{x}_d) - \mathbb{C}_0)[\boldsymbol{\varepsilon}^i(\boldsymbol{x}_d)], \quad \forall \boldsymbol{x}_d \in \Omega$
	ii)	$\hat{\boldsymbol{\tau}}^i = \mathrm{FFT}(\boldsymbol{\tau}^i)$
	iii)	$\hat{\boldsymbol{\varepsilon}}^{i+1}(\boldsymbol{\xi}_d) = -\hat{\mathbb{G}}_0(\boldsymbol{\xi}_d)[\hat{\boldsymbol{\tau}}^i(\boldsymbol{\xi}_d)],\, \forall \boldsymbol{\xi}_d \neq \mathbf{0},\, \hat{\boldsymbol{\varepsilon}}^{i+1}(\mathbf{0}) = \bar{\boldsymbol{\varepsilon}}$
	iv)	$\boldsymbol{\varepsilon}^{i+1} = \mathrm{FFT}^{-1}(\hat{\boldsymbol{\varepsilon}}^{i+1})$
	v)	convergence test: $\mathrm{res} = \frac{\lVert \boldsymbol{\varepsilon}^{i+1} - \boldsymbol{\varepsilon}^i \rVert_{L_2}}{\lVert \boldsymbol{\varepsilon}^1 - \boldsymbol{\varepsilon}^0 \rVert_{L_2}}$

Herein, $\boldsymbol{x}_d$ denotes coordinates of the voxels in real space, and $\boldsymbol{\xi}_d$ indicates the corresponding frequencies in Fourier space. Furthermore, the L^2-norm is defined as $\lVert (\cdot) \rVert_{L_2}^2 = \int_\Omega \lVert (\cdot) \rVert^2\, \mathrm{d}\Omega$.

Using the principle of equivalent strain energy

$$\frac{1}{2}\langle \boldsymbol{\varepsilon}_\alpha \cdot \mathbb{C}[\boldsymbol{\varepsilon}_\beta]\rangle = \frac{1}{2}\langle \boldsymbol{\varepsilon}_\alpha\rangle \cdot \bar{\mathbb{C}}[\langle \boldsymbol{\varepsilon}_\beta\rangle], \tag{5.38}$$

(Hill, 1963), the effective stiffness $\bar{\mathbb{C}}$ is calculated by solving the periodic boundary value problem given in Equation (5.34) for six linearly independent load cases $\boldsymbol{\varepsilon}_\alpha$ and $\boldsymbol{\varepsilon}_\beta$ with $\alpha, \beta \in \{1, \dots, 6\}$. The effective stiffness is then given by

$$\langle \mathbb{C}[\boldsymbol{\varepsilon}_\beta]\rangle = \bar{\mathbb{C}}[\langle \boldsymbol{\varepsilon}_\beta\rangle], \tag{5.39}$$

(Bishop and Hill, 1951a;b). By means of the commercial software tool GeoDict®, a voxel-based image obtained by a μCT scan can be directly imported as a *`.raw` file. After the segmentation process, further FFT-based analysis can be performed, and the local, as well as the overall material properties, are computed.

Chapter 6

Mean- and Full-Field Simulations for Long Fiber-Reinforced Polymers

6.1 Preliminaries

The mean- and full-field methods are widely used to compute the effective behavior of composite materials. In this chapter, the mean- and full-field simulation techniques as applied to different SMC composite samples are presented. The samples are extracted from individual composite plaques at different orientations. The production of the SMC plaques is described in Section 4.1.2. For the results presented in this chapter, the two plaques considered are denoted by 1D/1 and 1D/2. During the production of these plaques, the semi-finished product flows in the y-direction, cf. Figure 4.3a in Section 4.1.2. Additionally, a third plaque is considered, referred to as 2D, where the semi-finished product flows in the x- and y-directions during the production process, cf. Figure 4.3b. A µCT-scan is provided for each sample. The images of the samples of the plaques 1D/1, 1D/2, and 2D have a resolution of $5.5\ \mu m$, i.e. the edge length of each voxel amounts to $5.5\ \mu m$. Moreover, a higher resolution of $3\ \mu m$ is provided for the samples of plaque 1D/2.

To distinguish the scans at different resolutions, the samples of 1D/2 at higher resolutions are denoted as 1D/2 HR. The datasets used are summarized in Table 6.1.

	1D/1	1D/2, 1D/2 HR	2D
Elastic properties	HS two-step, FFT	HS two-step, FFT	HS two-step
DMA data	✓	✓	✓
Thermal properties	HS two-step, FFT	HS two-step, FFT	HS two-step
CTE data	✓	–	–

Table 6.1: Datasets used for the simulations and experimental investigations. Available datasets are indicated by "✓" whereas "–" denote non-available datasets.

The parameter set used for the numerical calculations is listed in Table 6.2. For the numerical simulations presented in this chapter, the fiber and matrix stiffnesses are assumed to be isotropic. Therefore, two engineering constants are used, namely, Young's modulus E and Poisson's ratio ν of the corresponding material. Analogously, the thermal expansion of the fiber and matrix materials is assumed to be isotropic as well and is given in terms of the scalar coefficients of thermal expansion (CTE) α_θ listed in Table 6.2. The values E_M and α_M of the matrix material are determined by means of DMA and CTE measurements presented in Sections 4.1.3 and 4.2.3. The nominal fiber volume content for the composite is predefined by the production process and amounts to 22 vol.%. However, Trauth and Weidenmann (2018) determined the real fiber volume content by thermogravimetric analyses for various samples from SMC plaques. According to their findings, the real fiber volume content is significantly higher than the nominal content. Thus, the simulations presented in the following sections, consider a fiber content of 25.6 vol.%, cf. Table 3 in Trauth and Weidenmann (2018). In addition, results of the mean-field calculations with respect to the theoretical fiber volume content of $c_{\mathrm{F}} = 0.22$ are provided in Appendix C.2.

Material	Parameters
Matrix	$E_{\mathrm{M}} = 3.4\,\mathrm{GPa}$, $\nu_{\mathrm{M}} = 0.385$, $\alpha_{\mathrm{M}} = 7.4 \times 10^{-5}/\mathrm{K}$, $c_{\mathrm{M}} = 0.744$
Fiber	$E_{\mathrm{F}} = 73.0\,\mathrm{GPa}$, $\nu_{\mathrm{F}} = 0.22$, $\alpha_{\mathrm{F}} = 0.5 \times 10^{-5}/\mathrm{K}$, $c_{\mathrm{F}} = 0.256$

Table 6.2: The parameter set for the UPPH matrix is determined by DMA and CTE measurements, cf. Section 4.1.3 and Section 4.2.3. The parameter values for the glass fiber material (E-glass fibers) are taken from Granta Design (2016).

The mean-field results are computed by the HS method presented in Section 5.1.2. This method consists of two steps, for which either the fiber or the matrix stiffness is used as reference stiffness in the first step, leading to the upper or lower bound, respectively. If the upper bound is considered in the first step, the overall result after the second step is briefly denoted by HS^{+}. By implication, if the lower bound is used in the first step, the overall result is referred to as HS^{-}. Additionally, a variable reference stiffness dependent on parameter k is introduced in the second step. In the following sections, this parameter k is varied between 0.0 and 1.0 in ten equidistant steps. Thus, the influence of the reference stiffness is investigated, and the smallest deviations from experimental data are sought. The fiber orientation tensors used for the orientation average within the HS mean-field homogenization are documented in Appendix A.3. The programming language Python (version 2.7.12) with the libraries NumPy and SciPy (Jones et al., 2001; Oliphant, 2006) is used for the implementation of the mean-field method.

The full-field results are computed by the fast Fourier transformation (FFT) implemented in the commercial software tool GeoDict®. The setting parameters for the full-field computations are documented in Appendix A.4.

The effective stiffness computed by the mean- and full-field simulations is illustrated by a contour plot of the effective Young's modulus given by $\bar{E}(\boldsymbol{d}) = (\boldsymbol{d} \otimes \boldsymbol{d} \cdot \bar{\mathbb{C}}^{-1}[\boldsymbol{d} \otimes \boldsymbol{d}])^{-1}$ (Böhlke and Brüggemann, 2001). The vectorial direction $\boldsymbol{d}$ is parametrized in spherical coordinates, considering a unit sphere with $\boldsymbol{d} = \boldsymbol{d}(\vartheta, \varphi)$ and $\|\boldsymbol{d}\| = 1$. The quantities ϑ and φ denote the azimuth and the polar angle, respectively. Moreover, the effective thermal expansion coefficient is presented by $\bar{\alpha}_\theta = \boldsymbol{d} \cdot \bar{\boldsymbol{\alpha}}_\theta[\boldsymbol{d}]$. In the subsequent sections, the contour plots of the effective Young's modulus and the effective CTE are depicted in the x-y-plane which corresponds to $\boldsymbol{d} = \boldsymbol{d}(\pi/2, \varphi)$. In this context, $\bar{E}(\boldsymbol{d})$ and $\bar{\alpha}_\theta(\boldsymbol{d})$ refer to $\boldsymbol{d} = \boldsymbol{d}(\pi/2, \varphi)$.

To present a short overview of the results provided in the following sections, the outline of this chapter is as follows. First, the effective elastic stiffness at $\theta = 20°\mathrm{C}$ is computed by the $\mathrm{HS}^{\pm}$ mean-field method. For different oriented samples of three plaques, the effective Young's moduli are compared to the corresponding experimental data obtained by DMA. Using the FFT-based full-field method, the synthetic microstructures are considered in what follows. The influence of the resolution of a µCT scan on the effective elastic properties is investigated. Subsequently, for real microstructures obtained by µCT scans, the effective elastic properties computed by the full-field method are compared to both mean-field results and experimental data. For the next investigations, the effective elastic material behavior with temperature-dependent elastic properties are computed using the mean-field method. The temperature-dependent effective Young's moduli are compared to the storage moduli from the temperature sweep tests presented in Section 4.1.3. Furthermore, the effective CTEs at $\theta = 20°\mathrm{C}$ are computed by the full- and mean-field methods. Additionally, the temperature-dependent CTE over the considered temperature range is predicted by the mean-field method and compared to the experimentally determined data.

6.2 Mean-Field Homogenization for DiCo Fiber-Reinforced UPPH at $\theta = 20°\mathrm{C}$

6.2.1 Effective elastic material properties

The average fiber orientation tensors within a volume element can be calculated from a μCT scan of a microstructure (Pinter et al., 2018). As discussed in Section 3.2, the fiber length is significantly larger than the thickness of a plaque or a sample. Thus, the contribution along the thickness direction, i.e. the z-direction, is neglected. As a consequence, deviations from the planar isotropic orientation tensor $\boldsymbol{N}^{\mathrm{pi}}$, given by

$$\boldsymbol{N}^{\mathrm{pi}} \,\widehat{=}\, \frac{1}{2}\begin{pmatrix} 1 & 0 & 0 \\ 0 & 1 & 0 \\ 0 & 0 & 0 \end{pmatrix}, \tag{6.1}$$

serve as a measure of anisotropy. For each sample with the fiber orientation tensor $\boldsymbol{N}_\alpha$, the relative error $a_\alpha^{\mathrm{pi}} = \|\boldsymbol{N}_\alpha - \boldsymbol{N}^{\mathrm{pi}}\| / \|\boldsymbol{N}^{\mathrm{pi}}\|$ is calculated, with $\alpha \in \{0°, 45°, 90°\}$ indicating the orientation of the sample in the cured plaque. The relative error a_α^{pi} quantifies the overall deviation of the averaged orientation of a sample from the planar isotropic reference orientation $\boldsymbol{N}^{\mathrm{pi}}$. Additionally, the eigenvalues for the corresponding orientation tensor are calculated. In this context, $\lambda_1 \geq \lambda_2 \geq \lambda_3$ holds. The third eigenvalue λ_3 can be calculated by means of the other two eigenvalues since $\mathrm{tr}(\boldsymbol{N}) = 1$ holds true. Thus, it is sufficient to plot the tuple $\{\lambda_1, \lambda_2\}$ in $\boldsymbol{R}^2$ to provide a graphical representation of the orientation tensors of the second-order. Characteristic eigenvalues are given in terms of isotropic, planar isotropic, and unidirectional microstructures, cf. Figure 3.4 for a graphical visualization. These characteristic eigenvalues define the edges of a triangle in the $\{\lambda_1, \lambda_2\}$ plane. The edge between the points PI and UD contains microstructures with perfectly planar distributed fibers corresponding to orientation tensors with $\lambda_3 = 0$. The location of the plotted tuples $\{\lambda_1, \lambda_2\}$ within this triangle specifies the main orientation tendency of a considered microstructure, cf. e.g. Figure 6.1b. In addition to the relative error a_α^{pi}, the depicted eigenvalue tuple resolves the deviation from the planar isotropic reference orientation $\boldsymbol{N}^{\mathrm{pi}}$ in more detail.

2D plaque. For samples from the 2D plaque, values of a_α^{pi} for the $\{0°, 45°, 90°\}$-oriented samples are shown in Figure 6.1a. The deviations of the $0°$ and $45°$ samples from the planar isotropic state are smaller than 10%. A relative error of nearly 15% is obtained between the average fiber orientation of the $90°$ sample and the planar isotropic state. In Figure 6.1b, the eigenvalues λ_1 and λ_2 for the three samples are shown. Both eigenvalues for the $0°$ and $45°$ samples assume the value of approximately 0.5, whereas a small deviation of the $90°$ sample of the planar isotropic orientation is present. However, the eigenvalue tuples are located close to the PI point. Thus, the three samples approximately exhibit a planar isotropic fiber orientation.

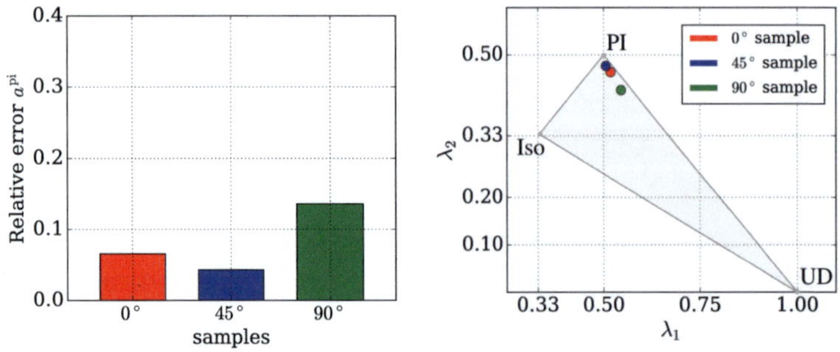

(a) (b)

Figure 6.1: a) Relative error $a_\alpha^{\mathrm{pi}} = \|\boldsymbol{N}_\alpha - \boldsymbol{N}^{\mathrm{pi}}\| / \|\boldsymbol{N}^{\mathrm{pi}}\|$ between the orientation tensors $\boldsymbol{N}_\alpha$ of $\alpha \in \{0°, 45°, 90°\}$ samples of the 2D plaque and the planar isotropic reference orientation tensor $\boldsymbol{N}^{\mathrm{pi}}$. b) Eigenvalues λ_1 and λ_2 of the orientation tenors, with $\lambda_1 \geq \lambda_2$. The abbreviations PI, Iso, and UD denote planar isotropic, isotropic, and unidirectional. All samples exhibit a pronounced planar isotropic microstructure.

In Figure 6.2, the measured elastic modulus and computational results of the mean-field simulations are shown in the x-y-plane, corresponding to $\bar{E}(\pi/2, \varphi)$. The effective Young's modulus for the HS$^+$ method is shown in the upper row for the three differently oriented samples. The corresponding results for the HS$^-$ method are depicted in the lower row. Both, the upper and the lower rows illustrate the effective material behavior, depending on the parameter k. The experimental results for the corresponding orientation are depicted by a red dot in each case. The effective material properties of the other plaques under consideration are presented analogously.

For the HS$^+$ method, the effective elastic modulus $\bar{E}(\boldsymbol{d})$ in the polar plane exhibits approximately the same quantitative and qualitative distribution in all the three samples. Moreover, for each sample, the angular distribution of $\bar{E}(\boldsymbol{d})$ exhibits a circular shape, predicting a distinct planar isotropic behavior. The influence of the parameter k on the effective elastic behavior is negligibly small for all three samples. Furthermore, the results of HS$^+$ are significantly stiffer than the experimental data.

In the HS$^-$ method, the distribution of $\bar{E}(\boldsymbol{d})$ slightly differs from one sample to another. The angular distribution of $\bar{E}(\boldsymbol{d})$ for each sample depends weakly on the polar angle φ. This indicates a slight anisotropy of the effective elastic behavior. The HS$^-$ results of all three samples depend on the parameter k. The values of $\bar{E}(\boldsymbol{d})$ vary between approximately $8\,\mathrm{GPa}$ and $12\,\mathrm{GPa}$, depending on the choice of k. Furthermore, the results of HS$^-$ are consistently smaller than or equal to the experimental data for all values of k. In particular, the computed effective elastic behavior exhibits small deviations from the experimental data for the $0°$- and $45°$-oriented samples with increasing values of k. Comparatively larger deviations are observed for the $90°$-oriented sample. For all three orientations, the HS$^-$ results are more compliant compared to the results from the HS$^+$ method. Since the HS$^+$ method is formulated in terms of the fiber stiffness $\mathbb{C}_F$ in the first step, consequently, the effective behavior is rather stiff. In contrast, the HS$^-$ method uses the softer matrix stiffness $\mathbb{C}_M$ in the first step, and thus, a reasonably more compliant effective behavior is obtained.

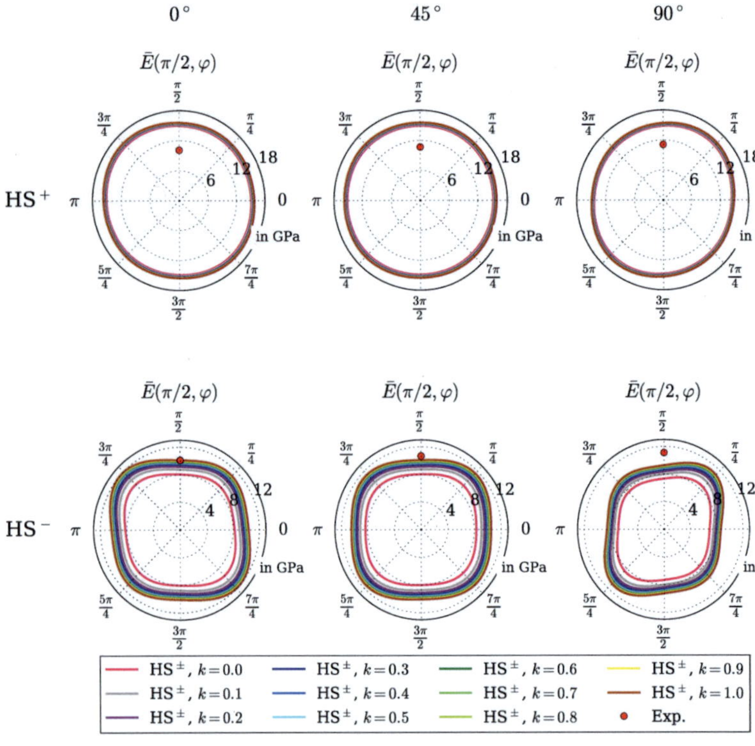

Figure 6.2: Experimental data are compared to results of the HS mean-field simulations for 0°, 45° and 90° samples of the 2D plaque. The HS$^+$ and HS$^-$ results are depicted in the upper and lower row, respectively. The parameter k of the variable reference stiffness varies. The experimental results are depicted by the red dots. The effective Young's modulus is illustrated in the x-y-plane, corresponding to $\bar{E}(\pi/2, \varphi)$. The simulation results are based on a μCT scan at a resolution of 5.5 μm. A negligible sensitivity of the HS$^+$ method on the variation of k is observed. Significant overestimation of the experimental results by the HS$^+$ method is given. The results of the HS$^-$ method are close to experimental data.

In order to quantify the deviations between the experimental data and the effective results obtained by the HS method, the relative error is computed depending on parameter k. Since the effective results from HS^+ exhibit a vanishing sensitivity with respect to k, only the relative error between experimental data and results from the HS^- method is shown in Figure 6.3. Irrespective of the considered sample's orientation, the relative error decreases monotonically with an increasing value of k. For $k = 1.0$, the relative error of the $90°$-oriented sample amounts to nearly $16\,\%$ whereas it is approximately $5\,\%$ for the $45°$-oriented sample. With respect to the $0°$-oriented sample, the deviation between the experiment and the simulation almost vanishes.

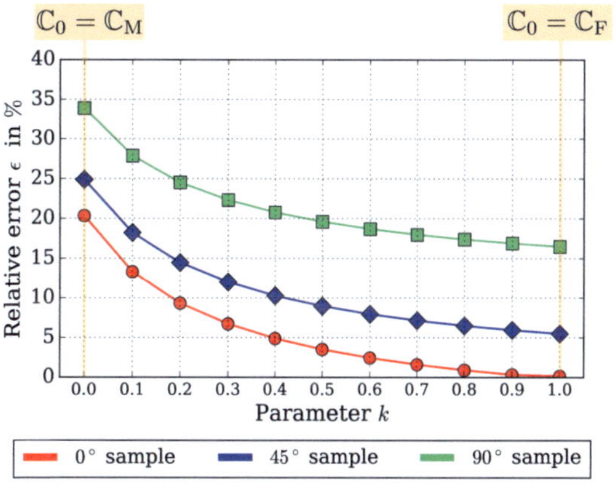

Figure 6.3: The relative error $\epsilon = |E_{sim}/E_{exp} - 1.0| \cdot 100\,\%$ between experimental data and simulations for HS^- is depicted for different values of k. Experimental data are obtained for $0°$, $45°$ and $90°$ samples of the 2D plaque. The relative error decreases monotonically for increasing values of k. Deviations increase with increasing angle between sample and load direction.

1D/1 plaque. The flow of the semi-finished product during the production process strongly influences the orientation of the fibers. This is illustrated by the orientation analysis, cf. Figure 6.4a. All the three samples exhibit deviations from the planar isotropic orientation state, quantified by the relative error a^{pi}. Herein, the 0°-oriented sample exhibits a deviation of nearly 30%. The smallest deviation is given for the 90° sample with $a^{\mathrm{pi}} = 18\%$. This is also illustrated in more detail in Figure 6.4b. The 0° and 45° samples exhibit a more pronounced anisotropic behavior compared to the 90° sample, as is shown by λ_1 and λ_2. The three points are located close to the PI-UD segment. Thus, the considered orientation is still significantly planar. However, the points are considerably shifted from the PI corner toward the UD corner, compared to Figure 6.1b. This illustrates the strong influence of the production process on the fiber orientation.

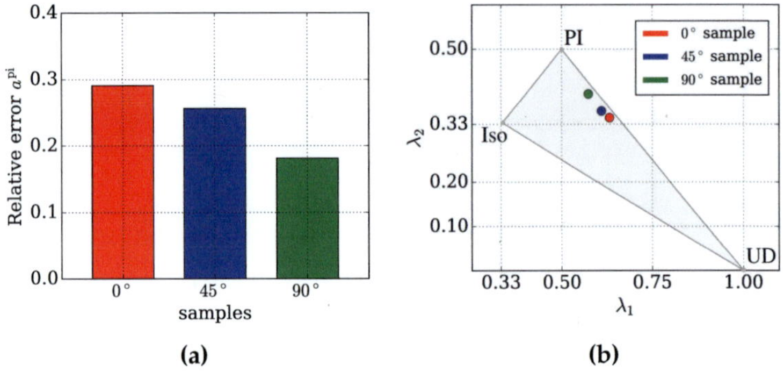

Figure 6.4: a) The relative error $a_\alpha^{\mathrm{pi}} = \|\boldsymbol{N}_\alpha - \boldsymbol{N}^{\mathrm{pi}}\| / \|\boldsymbol{N}^{\mathrm{pi}}\|$ between the orientation tensors $\boldsymbol{N}_\alpha$ of the $\{0°, 45°, 90°\}$ samples of the 1D/1 plaque and the planar isotropic reference orientation tensor $\boldsymbol{N}^{\mathrm{pi}}$. b) Eigenvalues λ_1 and λ_2 of the orientation tenors, with $\lambda_1 \geq \lambda_2$. The abbreviations PI, Iso, and UD denote planar isotropic, isotropic, and unidirectional. A pronounced deviation from the planar isotropic state is determined, especially for the 0° and 45° samples.

In Figure 6.5, the effective Young's modulus, computed by the HS method, is depicted for samples of the 1D/1 plaque.

The effective elastic behavior of the HS$^+$ method shows characteristic similarities to the previous findings. In particular, the computed effective Young's modulus exhibits a tendency toward an isotropic behavior with a negligible influence of parameter k on the effective behavior. Analogously, stiffer results are obtained compared to the given experimental data.

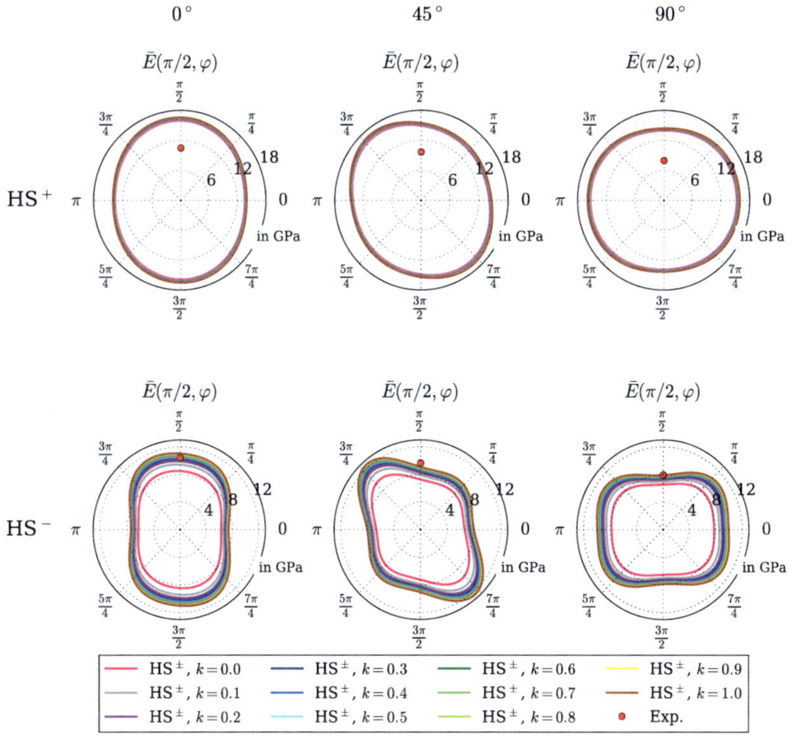

Figure 6.5: The experimental data are compared to the results of the mean-field simulation for different values of k for 0°, 45° and 90° samples of the 1D/1 plaque. The HS$^+$ and HS$^-$ results are depicted in the upper and lower row, respectively. The experimental results are depicted by red dots. The effective Young's modulus is illustrated in the x-y-plane, corresponding to $\bar{E}(\pi/2, \varphi)$. The simulation results are based on a µCT scan at a resolution of 5.5 µm. A significant overestimation of experimental results by the HS$^+$ method is observed. The HS$^-$ results are close to experimental data.

In the HS$^-$ method, $\bar{E}(\boldsymbol{d})$ for the three samples varies significantly from each other, which indicates the influence of the flow during the production process on the microstructure. The distribution of $\bar{E}(\boldsymbol{d})$ for a particular sample is strongly dependent on the polar angle φ, exhibiting a pronounced anisotropic behavior. Moreover, the effective results approach the experimental data with respect to increasing k. This holds true for all the three samples considered.

The relative error between the experiment and the effective Young's modulus of the HS$^-$ method is depicted in Figure 6.6. For the $45°$ and $90°$ samples, the error is a monotonically decreasing function of k. The $0°$-oriented sample attains the smallest relative error at $k = 0.4$. Higher values of k again lead to an increase in the relative error. At $k = 1.0$, the relative error amounts to less than $6\,\%$ in the case of the $45°$ sample and the $0°$ sample. The minimal relative error is about $1\,\%$ for the $90°$ sample.

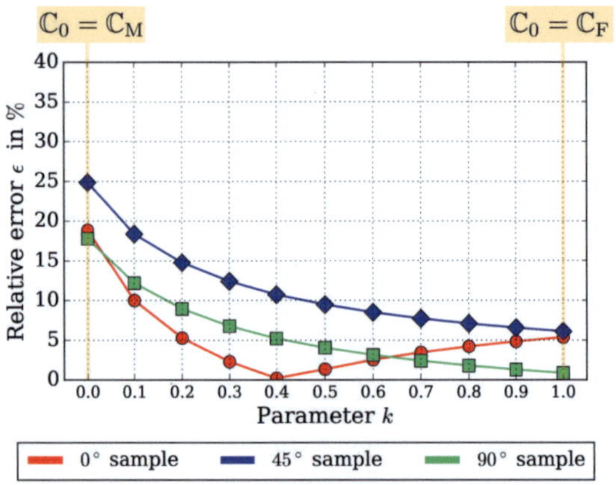

Figure 6.6: The relative error $\epsilon = |E_{\mathrm{sim}}/E_{\mathrm{exp}} - 1.0| \cdot 100\,\%$ between the experimental data and the simulations for HS$^-$ is depicted for different values of k. The experimental data are obtained for $0°$, $45°$ and $90°$ samples of the 1D/1 plaque. The relative error decreases monotonically for increasing values of k, considering the $45°$ and $90°$ sample. The deviation of the $0°$ sample exhibits a local minimum at $k = 0.4$.

1D/2 HR plaque. The third set of samples considered are taken from the 1D/2 plaque. Besides a scan with a resolution of $5\,\mu m$, an additional scan at a higher resolution of $3\,\mu m$ is provided for this plaque. Thereby, the influence of the resolution of the µCT scans on the predicted behavior is investigated. The results of the mean-field homogenization for the same samples with a resolution of $5.5\,\mu m$ are given in Appendix C.1 for completeness.

Consistent with the previous considerations, the relative error a^{pi} is depicted in Figure 6.7a. For this set, the largest deviation from the planar isotropic state is observed in the $45°$ sample with 30%, followed by the $90°$ sample. A relative error of nearly 18% is computed for the $0°$ sample. This is also illustrated in more detail in Figure 6.7b. The $45°$ and $90°$ samples exhibit a more pronounced anisotropic behavior compared to the $0°$ sample. Similar to the 1D/1 plaque, the eigenvalue tuples are located close to the PI-UD segment. The points are also shifted from the PI corner toward the UD corner. Again, this indicates the influence of the production process on the fiber orientation.

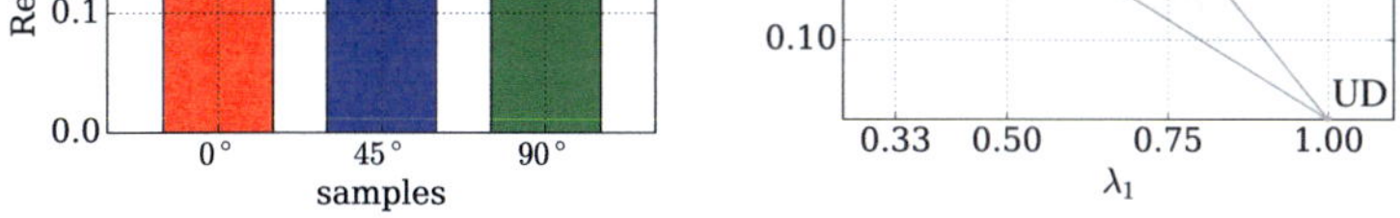

(a) (b)

Figure 6.7: a) The relative error $a^{\text{pi}}_{\alpha} = \|\boldsymbol{N}_{\alpha} - \boldsymbol{N}^{\text{pi}}\|/\|\boldsymbol{N}^{\text{pi}}\|$ between the orientation tensors $\boldsymbol{N}_{\alpha}$ of the $\{0°, 45°, 90°\}$ samples of the 1D/2 HR plaque and the planar isotropic orientation tensor $\boldsymbol{N}^{\text{pi}}$. b) Eigenvalues λ_1 and λ_2 of the orientation tenors, with $\lambda_1 \geq \lambda_2$. The abbreviations PI, Iso, and UD denote planar isotropic, isotropic, and unidirectional. A pronounced deviation from the planar isotropic state is determined, especially for the $45°$ and $90°$ samples.

The results from the HS$^+$ method for the samples of the 1D/2 HR plaque are shown in Figure 6.8.

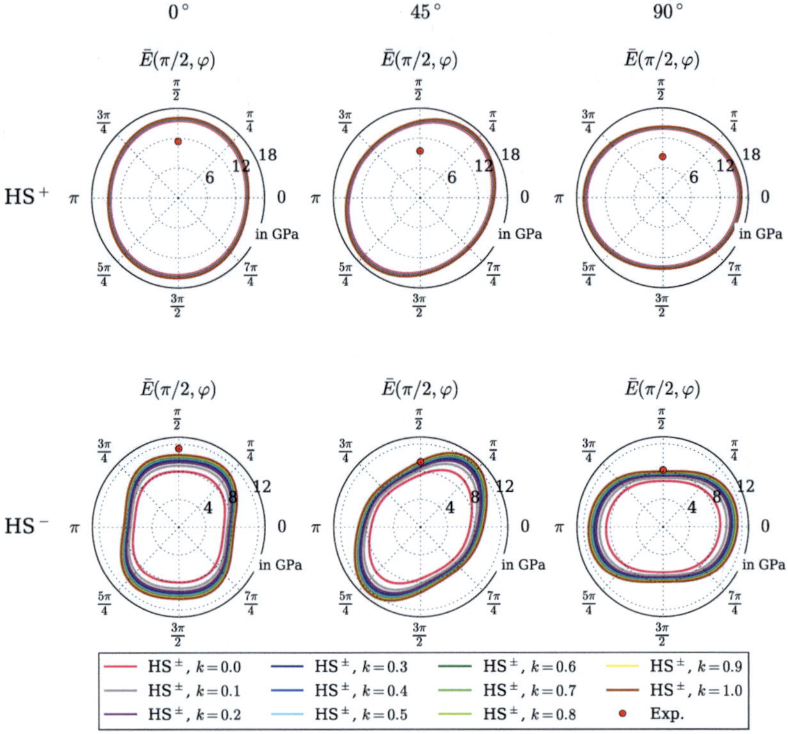

Figure 6.8: The experimental data are compared to the results of the HS mean-field simulations for different values of k for 0°, 45° and 90° samples of the 1D/2 HR plaque. The HS$^+$ and HS$^-$ results are depicted in upper and lower row, respectively. The experimental results are depicted by red dots. The effective Young's modulus is illustrated in the x-y-plane, corresponding to $\bar{E}(\pi/2, \varphi)$. The simulation results are based on a µCT scan at a resolution of 3.0 µm. A significant overestimation of the experimental results by the HS$^+$ method is observed whereas results of the HS$^-$ method are close to experimental data.

Similar to the foregoing results depicted in Figures 6.2 and 6.5, the effective elastic results from the HS$^+$ method exhibit a pronounced isotropic behavior. Moreover, the results are stiffer than the experimental data, and parameter k has nearly no influence on the material behavior. The distribution of $\bar{E}(\boldsymbol{d})$ from HS$^-$ for the different samples are rather ellipsoidal-shaped for varying φ. This is in contrast to the previous results of the 1D/1 plaque where a distinct anisotropic distribution of $\bar{E}(\boldsymbol{d})$ on the polar angle φ is observed. Consistent with the results of the samples from the other plaques, the effective elastic behavior depends on parameter k and is more compliant compared to experiments and the corresponding HS$^+$ results.

The relative error between the numerical results of the HS$^-$ method and the experimental data is depicted in Figure 6.9.

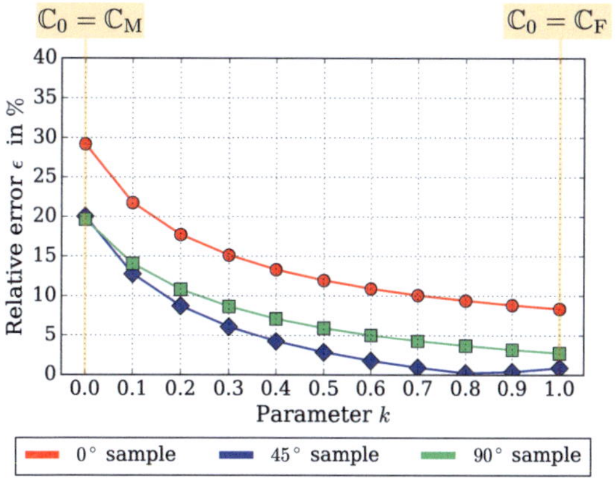

Figure 6.9: The relative error $\epsilon = |E_{\text{sim}}/E_{\text{exp}} - 1.0| \cdot 100\,\%$ between the experimental data and the simulations of the HS$^-$ method for different values of k. The experimental data are with respect to the 0°, 45°, and 90° samples of the 1D/2 HR plaque. The relative error monotonically decreases for increasing values of k regarding the 45° and 90° sample. The deviation of the 45° sample exhibits a local minimum for $k = 0.8$.

The error decreases monotonically with increasing k values. The minimal deviations from the corresponding experimental data are given for $k = 1.0$ regarding the $0°$- and $90°$-oriented sample. In terms of the $45°$-oriented sample, the smallest deviation is obtained for $k = 0.8$. At $k = 1.0$, the relative error amounts less than $10\,\%$ for $0°$ and less than $5\,\%$ for $45°$ and $90°$.

6.2.2 Summary of results

The effective elastic moduli at $20°\mathrm{C}$ obtained by the HS^+ method for all cases are considered isotropic. Moreover, the results obtained are generally stiffer than those obtained in the experiments. Müller (2016) observed the same tendencies for an alternative derivation of an HS two-step bounding method. This behavior is caused by the use of the fiber stiffness within the first step of the homogenization procedure. Since the phase contrast $E_\mathrm{F}/E_\mathrm{M}$ for the SMC composite considered amounts to nearly 21, the elastic fiber properties are significantly stiffer than those of the matrix material. Furthermore, only a slight dependence on the parameter k is observed for the HS^+ method. An essential disadvantage of the HS^+ method is reflected in the approximately similar predicted effective behavior, irrespective of the sample considered. Thus, the effective elastic material properties computed by HS^+ are less affected by the microstructure.

In contrast, the HS^- method exhibits an anisotropic effective stiffness throughout all plaques considered. Generally, more compliant results are obtained compared to the experimental data and the HS^- method. There is a pronounced dependency of the homogenization results on the parameter k. The largest deviation from the experimental data is present for $k = 0.0$, which corresponds to choosing $\mathbb{C}_0 = \mathbb{C}_\mathrm{M}$ in the second step. For increasing values of the parameter k, the deviation decreases and a reasonable fit to the experimental data is achieved. Except in two cases, the smallest deviation reached is for $k = 1.0$. Additionally, a dataset with a higher resolution is considered. In this case, a more ellipsoidal-shaped distribution of $\bar{E}(\boldsymbol{d})$ is obtained compared to the coarser resolution. However, for k reaching 1.0, both resolutions yield results that exhibit minor deviations compared to the corresponding experimental data. Since the relative deviation is only given with respect to one specific point, the full angular distribution of $\bar{E}(\boldsymbol{d})$ cannot be validated by the experimental data available. For this purpose, the full-field homogenization results are investigated in the next section.

6.3 Full-Field Homogenization for DiCo Fiber-Reinforced UPPH at $\theta = 20°C$

6.3.1 Preliminary studies on synthetic fiber structures

The FFT-based full-field method is directly applied to the segmented microstructure obtained by a µCT scan. The microstructure of an SMC composite is governed by bundled fibers, cf. Section 3.2. Separating the individual fibers in these closely packed bundles poses a significant challenge for image reconstruction. To investigate the influence of the resolution of the µCT scan on the effective elastic properties in detail, a resolution study was performed in a first step. To this end, volume elements with generated microstructures are considered. For this purpose, the synthetic fibers are modeled as straight fibers exhibiting a fiber diameter of $d_F = 10\,\mu m$ and an aspect ratio of 100. The fiber volume content of the synthetic microstructures is set to $27\,\text{vol.}\%$. As an average orientation, the second-order orientation tensor of the $90°$-oriented sample of the dataset 1D/2 is chosen. The volume elements at different resolutions are generated by an algorithm described by Schneider (2017b). Table 6.3 lists the considered resolutions as well as the dimensions of the volume elements. Based on a given volume element, a subvolume at different resolutions is considered as illustrated in Figure 6.10 for the subsequent resolution study.

The commercial software tool GeoDict® (Math2Market, 2017) is used for the FFT-based full-field homogenization of the SMC material considered. The settings for the computations are documented in Appendix A.4.

Resolution	Dimensions
1.5 μm	846 × 846 × 216 voxels
2.0 μm	640 × 640 × 160 voxels
3.0 μm	432 × 432 × 106 voxels
4.0 μm	320 × 320 × 80 voxels
5.0 μm	256 × 256 × 64 voxels

Table 6.3: Resolution and dimensions of the synthetic volume elements generated.

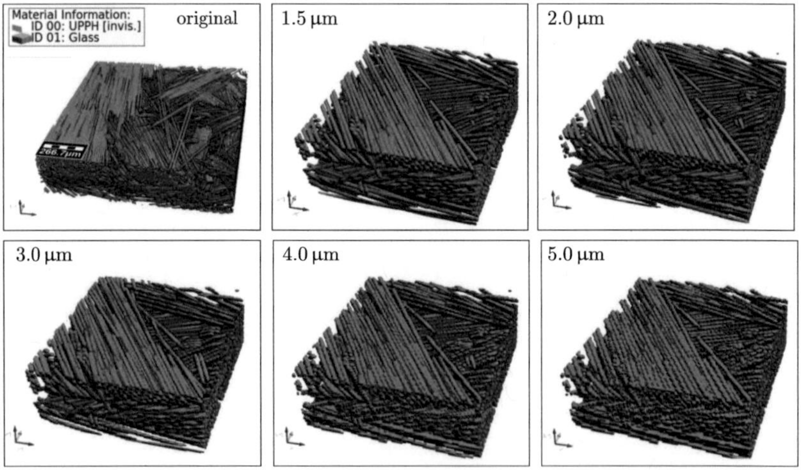

Figure 6.10: Synthetic microstructures for different resolutions.

Figure 6.11a shows the effective Young's modulus for the FFT solution for the corresponding resolutions.

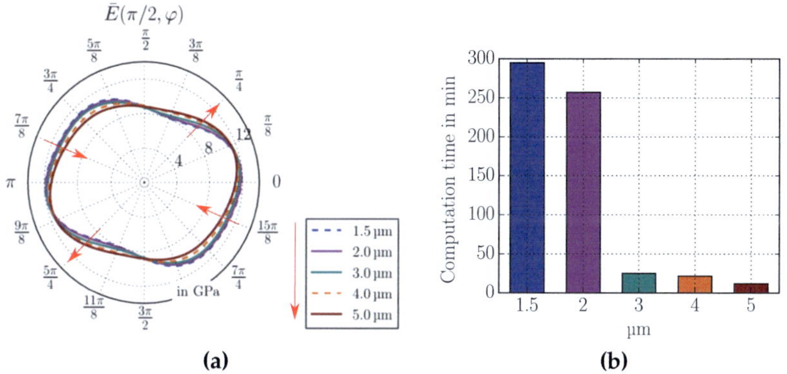

Figure 6.11: a) FFT-based solutions for microstructures at different resolutions. The red arrows indicate the tendencies of the results for decreasing resolution. b) Computation times for the corresponding simulations.

For a high resolution of $1.5\,\mu\text{m}$, an anisotropic distribution of $\bar{E}(\boldsymbol{d})$ is obtained. For coarser resolutions, the contour plot turns into a rather ellipsoidal-shaped distribution. In the ranges of $\varphi \in [\pi/8, \pi/2]$ and $\varphi \in [9\pi/8, 3\pi/2]$, the elastic moduli predicted for the resolutions of $5.5\,\mu\text{m}$ and $4\,\mu\text{m}$ are stiffer and become more compliant with increasing resolutions. In contrast, in the ranges of $\varphi \in [\pi/2, 9\pi/8]$ and $\varphi \in [3\pi/2, \pi/8]$, the observed behavior is reversed. Since the space between the single fibers within a fiber bundle is fairly small, only few voxels are present between the single fibers in the case of the high resolution of $1.5\,\mu\text{m}$ or $2\,\mu\text{m}$. With a coarser resolution, the voxel size becomes larger than the gaps between the fibers. Thus, with coarser resolutions, fibers appear connected and form bundles. This is illustrated by the ellipsoidal shape of $\bar{E}(\boldsymbol{d})$ in coarser resolutions compared to the higher resolutions. A pronounced ellipsoidal shape of $\bar{E}(\boldsymbol{d})$ is obtained for resolutions with voxel edge lengths greater than $4\,\mu\text{m}$. The maximum deviation between the highest and the coarsest resolution considered is nearly 17 %. According to expectations, a higher resolved microstructure results in longer calculation times due to the increase in the number of voxels. This is illustrated in Figure 6.11b. The results for resolutions coarser than 3 µm exhibit a substantial reduction in calculation time. Coarsening the resolution by a factor of 2 leads to a savings in computation time of factor 8 since the voxel size decreases nearly by a factor of 8. For instance, when using a resolution of 2 µm instead of 4 µm, the results obtained in this particular case are improved by a maximum of 10%, but the computation time increases significantly. Moreover, for real microstructures, increasing the resolution either leads to a decrease of the geometrical dimensions of the volume element considered or to a high memory demand. Taking this and a reasonable computation time into account, resolutions of 3 µm and coarser are appropriate.

6.3.2 Effective elastic material properties

Original µCT data. The effective elastic material behavior is computed by an FFT-based homogenization for three differently oriented samples of the dataset 1D/2. To keep the computation effort to a minimum, a resolution of 5.5 µm is used. As input parameters for the matrix as well as for the fiber material, the isotropic material constants used are according to Table 6.2. The µCT data are directly imported, either as *.RAW file or as stacked *.TIF files.

Figure 6.12 shows the effective Young's modulus $\bar{E}(\boldsymbol{d})$. In addition to the numerical results by the FFT method, the mean-field solutions are shown for the HS^- method with $k = 1.0$ and the corresponding experimental data. A scan at a higher resolution of 3 µm is additionally used for the mean-field method. The results of the mean-field method associated to the scan with a 3 µm resolution are denoted by "HR".

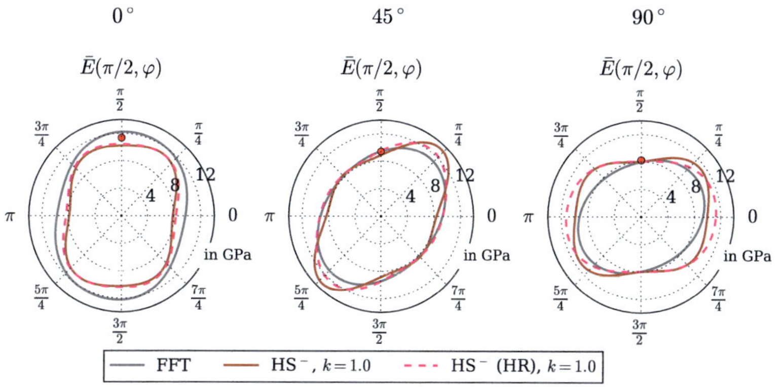

Figure 6.12: Full-field simulation results compared to the experimental data as well as to the mean-field simulation results at resolutions of 5.5 µm and 3 µm.

The left picture in Figure 6.12 shows the results of the $0°$ sample. The mean-field results for the distribution of $\bar{E}(\boldsymbol{d})$ reveal no significant differences between the results of both resolutions. Deviations are observed in the FFT results with respect to the angular distribution of $\bar{E}(\boldsymbol{d})$ for varying φ values. Furthermore, the $\bar{E}(\boldsymbol{d})$ values are stiffer in the FFT method compared to the mean-field solutions. Moreover, the predicted result of the FFT method slightly overestimates the experimental data. However, the deviation between the FFT simulation result and the experimental data is small. For the $45°$-oriented sample, slight deviations between the HS$^-$ method are observed for the two datasets regarding the angular distribution of $\bar{E}(\boldsymbol{d})$. Furthermore, the mean-field method exhibits a higher predicted stiffness compared to the FFT results. The distribution of $\bar{E}(\boldsymbol{d})$ in the FFT method is in line with that in the HS$^-$ (HR) method. Indeed, both methods predict a rather ellipsoidal-shaped distribution of $\bar{E}(\boldsymbol{d})$. For φ values close to the loading direction, i.e. $\pi/2$, all considered methods nearly coincide. In the case of the $90°$ sample, all numerical results exhibit deviations in the ranges of $\varphi \in [3\pi/4, 5\pi/4]$ and $\varphi \in [7\pi/4, \pi/4]$. However, a good agreement between all methods and the experimental data is given in loading direction.

The relative error ϵ of the simulation results with respect to the experimental data of the three orientations is depicted in Figure 6.13. All the three simulation results show the largest relative error in the case of the $0°$ sample. The FFT result exhibits the smallest deviations among the computations considered. The application of the methods to the $45°$ sample shows an almost perfect fit for the high-resolution dataset and the FFT method. The HS$^-$ results for the coarser dataset exhibit comparatively larger deviations. However, the relative error of $\epsilon \approx 6\,\%$ is sufficiently small. In the case of the $90°$ sample, all numerical results have nearly the same relative error value. Herein, the relative error is less than $3\,\%$ for all three computations.

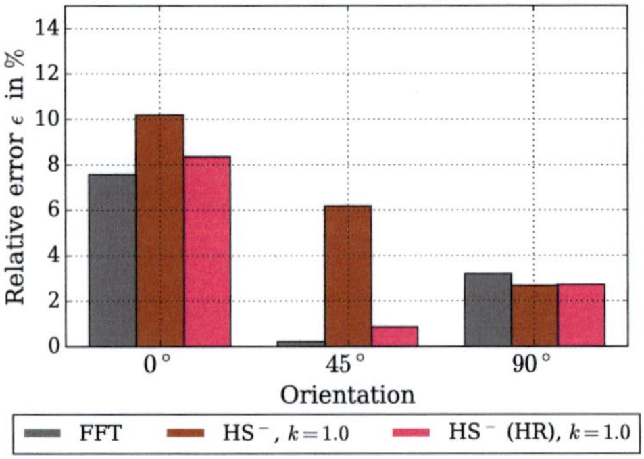

Figure 6.13: The relative error $\epsilon = |E_{\mathrm{sim}}/E_{\mathrm{exp}} - 1.0| \cdot 100\,\%$ between the experimental data and the simulation results for the FFT-based homogenization and the HS$^-$ method at two resolutions, $5.5\,\mu\mathrm{m}$ and $3\,\mu\mathrm{m}$. Samples of the 1D/2 plaque are considered.

Processed µCT data. For a dataset of nearly $1800 \times 1900 \times 560$ voxels with an edge length of $5.5\,\mu\mathrm{m}$, the FFT solution[1] takes about 3-4 days,

[1] Computations were performed on a 64-bit workstation system with a AMD Opteron^{TM} Processor (6238×24), and 128 GB RAM.

whereas the mean-field solution is obtained in a few milliseconds. To compare the mean-field results in detail with the results obtained by the FFT method, shorter computation times for the FFT method are desirable. To reducing the computation time while maintaining the accuracy of the results, a pre-processing step is employed. Based on the processing modules offered by the GeoDict® software tool, the geometry of the considered volume element can be reduced. By shrinking the volume data, the microstructure is made coarser. Since the fibers are strongly bundled after this pre-processing step, transversely isotropic material properties are assigned to the bundles. The material and setting parameters are given in Appendix A.4. An extended investigation regarding the pre-processing of CT data and microstructures with bundled fibers is given in Sliseris et al. (2014).

The reconstructed µCT scan of the 90°-oriented sample of the 1D/2 plaque is considered in the following. The original structure, as well as the processed structure, are shown in Figure 6.14. On the left-hand side, a 3-D and 2-D reconstruction of the µCT data is shown. The fibers within the original microstructure are poorly resolved and gaps within the fiber filaments are present. The pre-processing step causes the fibers to cluster into bundles. The gaps between the fibers in the left image are refilled, and thicker fiber bundles are obtained. Using the processed image, the microstructure is shrunk by the factors 2 and 4, respectively. Table 6.4 lists the dimensions of the volume elements and the calculation time for the corresponding microstructure.

	Dimensions	Shrink factor	Comp. time
FFT, original,	$1821 \times 1926 \times 558$ voxels	[−]	≈ 90h
FFT, shrink factor 2	$900 \times 960 \times 256$ voxels	2	≈ 5h
FFT, shrink factor 4	$455 \times 481 \times 139$ voxels	4	< 1h

Table 6.4: Dimensions, shrink factors, and corresponding computation times of the original and the reduced microstructures.

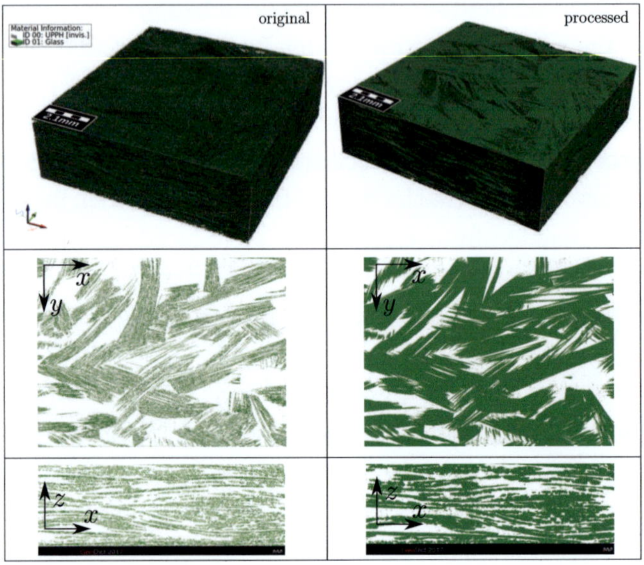

Figure 6.14: Imported original and processed microstructure of a μCT scan of the 90°-oriented sample of the 1D/2 plaque.

By reducing the geometry dimensions by a factor of 2, and thus the volume considered by a factor of 8, the computation time is already significantly reduced. The effective elastic properties are computed for the three microstructures and depicted in Figure 6.15.

Maximum deviations are observed in the range of $10 - 13\,\%$ between the three solutions. The simulation with a shrink factor of 2 results in slightly stiffer values of $\bar{E}(\boldsymbol{d})$ whereas the results with a shrink factor of 4 are more compliant. The effective Young's modulus for the original microstructure is distributed between the solutions for the shrink factors 2 and 4. However, the deviations from the experimental data are less than 6%.

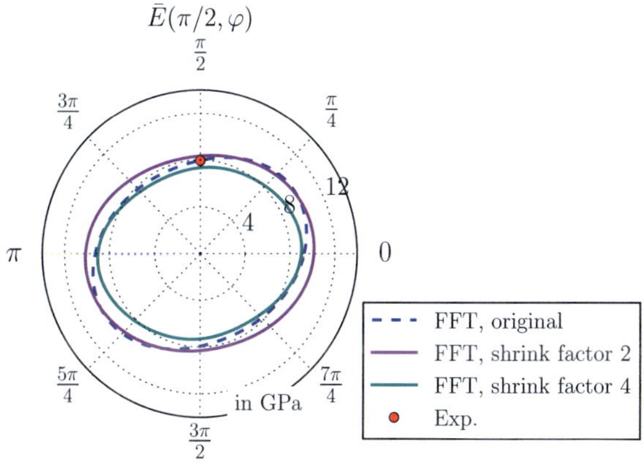

Figure 6.15: FFT-based results for the original and the processed microstructure. Shrink factors of 2 and 4 are used to reduced the volume element. The results are compared to experimental data for the 90°-oriented sample of the 1D/2 plaque.

6.3.3 Summary of results

Motivated by the studies on synthetic microstructures, a compromise has to be found between accuracy and computation time. When using a rather coarse resolution of $5.5\,\mu m$ (Pinter, 2018), the question arises of how well the numerical results will fit the experimental data. The foregoing findings indeed reveal a good agreement of the mean- and the full-field simulation results with the experimental data. These deviations do not exceed the experimental data by $10\,\%$. Thus, the presented methods are reliable regarding the prediction of the effective elastic material behavior. However, it has to be considered that this comparison is only given for one single point in the x-y-plane. Due to the lack of experimental data in different directions of one sample, the distribution of $\bar{E}(\boldsymbol{d})$ can only be compared for different simulation methods. The effective Young's modulus computed using the HS$^-$ mean-field method reveals deviations from the FFT-based full-field results with respect to the angular distribution of $\bar{E}(\boldsymbol{d})$.

The FFT results exhibit a rather ellipsoidal-shaped distribution whereas the mean-field results show a slightly more anisotropic behavior. Since the mean-field method uses orientation tensors as microstructural information, a different input is given compared to that used in the full-field method. This difference has an impact on $\bar{E}(\boldsymbol{d})$ and presumably causes the deviations in the contour of the plot. Additionally, the microstructure of the SMC composite is governed by bundled fibers that affect the overall behavior. The fiber bundles are not yet considered in the mean-field method and are poorly resolved in the full-field method. This might additionally contribute to the deviations observed.

The resolution study for a synthetic microstructure revealed that coarsening the resolution by a factor of 2 significantly reduces the computation time. Compared to the mean-field method, the FFT-based homogenization of microstructures of nearly 2.0×10^9 voxels requires extended computation times and considerable computation resources such as RAM (a minimum of $100\,\mathrm{GB}$). Motivated by the μCT microstructures processed, it is recommended to generate realistic virtual structures. This, however, requires an intensive study of the underlying real microstructures and a further consideration of the fiber bundles.

6.4 Temperature-Dependent Effective Elastic Material Properties

6.4.1 Mean-field homogenization results

The pure and fiber-reinforced SMC material was characterized experimentally in Section 4.1.3. The temperature strongly influences the material properties of the matrix and the composite. Thus, temperature-dependent material parameters need to be considered within the homogenization method. To investigate the temperature-dependent effective elastic material properties, the effective stiffness is computed as temperature-dependent function. Computing the effective stiffness for the discrete temperatures considered in the experiments results in a large number of computations. Due to the numerical efficiency, only mean-field but no full-field results are considered in the following investigations.

Young's modulus of the matrix material $E_{\mathrm{M}}(\theta)$ is determined as a function of temperature by conducting DMA tests. Due to the lack of experimental data for the temperature-dependent Poisson's ratio of the matrix material, it is modeled as temperature-independent. Moreover, Young's modulus and Poisson's ratio of the fiber material are assumed to be temperature-independent, as well.

The effective temperature-dependent stiffness $\bar{\mathbb{C}}(\theta)$ in terms of $E_{\mathrm{M}}(\theta)$ is calculated by means of the HS$^-$ method with $k = 1.0$. The temperature-dependent effective Young's modulus $\bar{E}(\theta, \boldsymbol{d})$ is determined and evaluated for $\varphi = \pi/2$ and $\vartheta = \pi/2$, corresponding to the tension direction of the corresponding sample.

Figure 6.16 shows the simulation results and the experimental data for the three samples of the 1D/2 plaque for the same temperature range considered in Section 4.1.3. The experimental data are depicted by a dotted line, whereas the simulation results are represented by solid lines. For the $0°$ sample, the numerical results are more compliant compared to the distribution of the experimental storage modulus. This deviation remains nearly constant for increasing temperatures of up to $150°\mathrm{C}$. In Section 6.2.1 a slight deviation between the simulation and the experimental data was already observed at $20°\mathrm{C}$. In the temperature range of $100°\mathrm{C}$ and close to the glass transition temperature at $150°\mathrm{C}$, the measured behavior of the composite becomes nonlinear. Since the storage modulus of the matrix material nearly vanishes above $150°\mathrm{C}$, cf. Section 4.1.3, the effective elastic stiffness is only affected by the temperature-independent fiber material. This behavior is not captured well by the mean-field method. However, for common applications of SMC composites, a temperature of 100°C is generally not exceeded. Hence, the results of the HS$^-$ method are in good agreement with experimental data for reasonable temperature ranges up to 100°C.

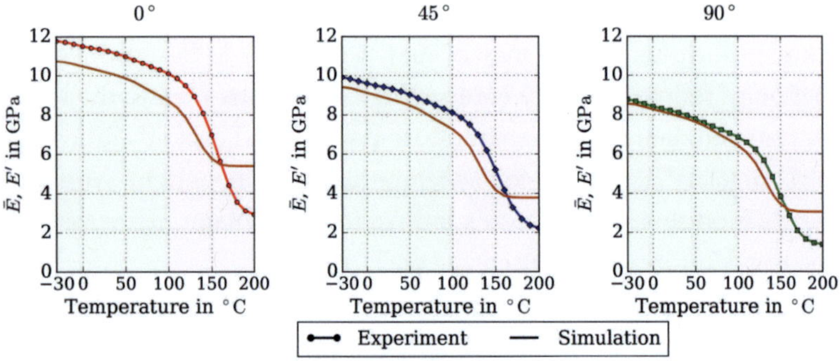

Figure 6.16: Comparison of the effective, temperature-dependent Young's modulus, obtained by the HS$^-$ method, with the experimental data of the 0°, 45°, and 90° samples. The simulations are performed for $k = 1$. The experimental data are obtained from the 1D/2 plaque. The deviation between the simulation and the experiment decreases with the increasing angle between the orientation of the sample in the cured plaque and the flow direction. The temperature range for common applications is highlighted in green.

Within the temperature range of −30°C to 100°C, the predicted effective Young's modulus for the 45°-oriented sample exhibits small deviations from the measured storage modulus. A similar behavior is observed for the 90°-oriented sample. In particular, the simulation results exhibit excellent agreements with the experiments up to 100°C. Generally, the deviation between the effective and the storage modulus decreases for an increasing inclination angle between the main fiber orientation and the loading direction. Similar to the findings for the 0° sample, the behavior above 150°C is not captured.

Figure 6.17 quantifies the relative error between the predicted and the measured elastic moduli depending on the temperature. For all orientations considered, the relative error increases for temperatures above 100°C. For temperatures below 100°C, the deviations are in the range of $\epsilon \approx 2\,\%$ (90° sample) to $\epsilon \approx 14\,\%$ (0° sample).

For an increasing temperature and, thus, a decreasing storage modulus of the matrix, the relative error increases substantially. In the region of $\theta > 100$°C, the error ϵ increases to a local maximum. Thereafter, it decreases again and shortly after 150°C, it finally blows up. The decreasing storage modulus correlates with an increasing viscous behavior of the material, which is not captured by the mean-field model. Thus, the observed increase of the relative error with an increasing temperature is expected. Moreover, the physical processes around the glass transition temperature are not accounted for by the mean-field model as well. Consequently, the transition region around the glass transition temperature θ_g is not captured well by the model. The behavior for $\theta \leq 100$°C, however, exhibits small deviations from the experimental data for all three samples.

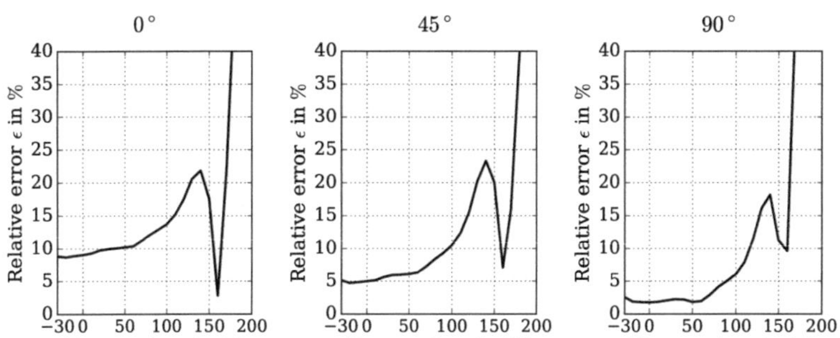

Figure 6.17: The relative error $\epsilon = |E_{\text{sim}}/E_{\text{exp}} - 1.0| \cdot 100\,\%$ between the experimental data and the effective, temperature-dependent Young's modulus, obtained by the HS$^-$ method for varying k. The experimental data are with respect to the 0°, 45°, and 90° samples of the 1D/2 plaque. The singularity of the error is for values of θ close to θ_g. The temperature range for common applications is highlighted in green.

6.4.2 Summary of results

As discussed in Section 4.1.3, the matrix behavior is strongly influenced by the temperature load. Thus, inevitably, a material model has to account for this temperature dependency. In the temperature range $\theta \in [-30, 100]^\circ\mathrm{C}$, the predicted effective elastic moduli $\bar{E}(\theta)$ of all three samples agree well with the corresponding experimentally determined storage moduli. The deviation between the simulation results and the experimental data decreases with an increase in the angle between the main fiber orientation and the load direction. In particular, the fibers of the 90°-oriented sample are preferably oriented transversely to the direction of the tension load. This leads to a matrix-dominated behavior which is captured well by the HS^- mean-field model. For temperatures around $100^\circ\mathrm{C}$ and close to the glass transition temperature at $150^\circ\mathrm{C}$, a nonlinear decreasing behavior of the storage moduli is measured. These effects are not accounted for by the mean-field model. Consequently, the deviations between the experiments and the simulation results increase in this temperature region. However, this temperature region is out of the reasonable range of applications. Regarding the reasonable temperature range, the results of the developed mean-field method are in good agreement with the experimental data.

6.5 Effective Thermal Properties

6.5.1 Mean-field homogenization results for $\theta = 20^\circ\mathrm{C}$

The effective coefficient of thermal expansion (CTE) of the SMC composite is calculated by the HS^+ and HS^- methods as stated in Equation (5.28). For this purpose, the CTE values of the fiber and the matrix material are used according to Table 6.2. Applying HS^+ and HS^- to the three differently oriented samples of the plaque 1D/1, the effective CTE $\bar{\alpha}_\theta(\boldsymbol{d})$ is determined and shown in Figure 6.18. As with the results previously presented, the results obtained by the HS$^+$ method are depicted in the upper row, while the HS$^-$ results are given in the lower row. The values of parameter k vary between 0.0 and 1.0 in ten equidistant steps.

Regarding the HS$^+$ method, an ellipsoidal-shaped distribution of $\bar{\alpha}_\theta(\boldsymbol{d})$ is observed for the three samples. A dependency of the simulation results on the parameter k is present. Choosing values smaller than 0.3 for k yields a more pronounced ellipsoidal-shaped distribution of $\bar{\alpha}_\theta(\boldsymbol{d})$. For increasing values of k, the shape tends toward a planar isotropic contour. Compared to the corresponding experimental data, the HS$^+$ method generally overestimates the measured material data in the case of the $0°$ and the $45°$ sample. For the $90°$-oriented sample, the calculated effective CTE is close to the experimental CTE value and exhibits vanishing deviations.

For the HS$^-$ method, an ellipsoidal-shaped distribution of $\bar{\alpha}_\theta(\boldsymbol{d})$ is obtained as well. The impact of small values of k on the distribution $\bar{\alpha}_\theta(\boldsymbol{d})$ is higher compared to the HS$^+$ method. Values of $k \leq 0.3$ lead to a pronounced ellipsoidal-shaped distribution of $\bar{\alpha}_\theta(\boldsymbol{d})$, whereas $k \geq 0.3$ yield a more planar isotropic behavior of the results. Irrespective of the orientation considered, the effective CTE overestimates the experimental CTE values.

Comparing the results from HS$^+$ and HS$^-$ with the experimental CTE, the smallest deviations are computed for the results obtained by the HS$^+$ method with $k = 1.0$. The HS$^-$ method leads to results that overestimate the experimental data due to the high CTE of the matrix in the first homogenization step. In contrast, the HS$^+$ method uses the CTE of the fibers in the first step which is significantly lower compared to CTE of the matrix. Thus, smaller $\bar{\boldsymbol{\alpha}}_\theta$ values are expected. Consequently, the HS$^+$ with $k = 1.0$ is used for the subsequent computations.

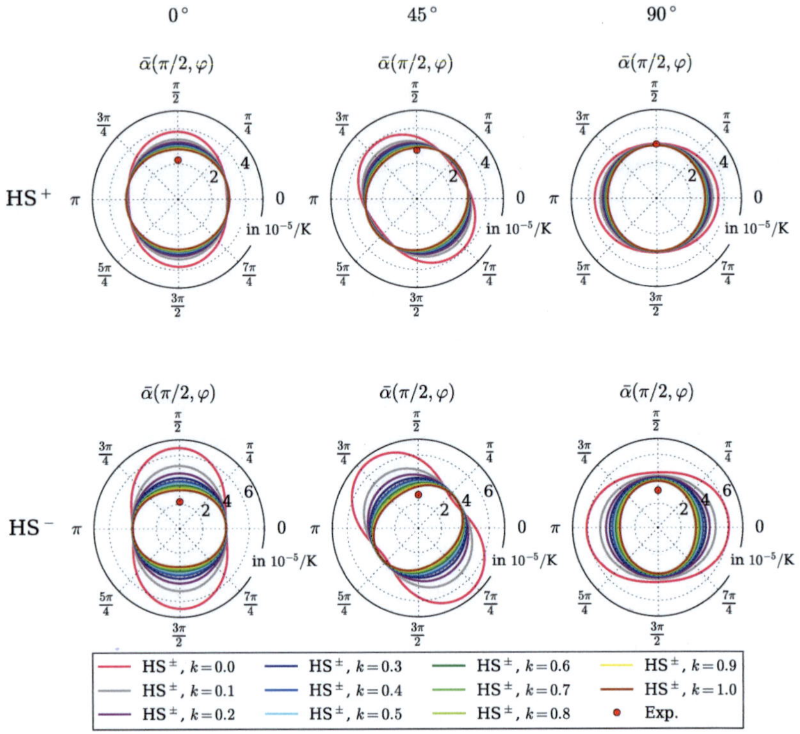

Figure 6.18: Comparison of the experimental data with the results of the mean-field simulations for different values of k regarding the 0°, 45° and, 90° samples of the 1D/1 plaque. The HS$^+$ and HS$^-$ results are depicted in the upper and lower rows, respectively. The experimental results are depicted by red dots. The illustration of the effective CTE in the x-y-plane corresponds to $\bar{\alpha}_\theta(\pi/2, \varphi)$. The simulation results are based on a µCT scan at a resolution of 5.0 µm. There is a slight overestimation of experimental results by the HS$^-$ method. Good agreements with experimental data are achieved for the HS$^+$ method for the 45° and 90° samples.

6.5.2 Full-field homogenization results for $\theta = 20°\mathrm{C}$

By means of the software tool GeoDict®, the effective CTE is computed by applying the FFT method to the µCT data for the three different samples of the 1D/1 plaque. In Figure 6.19, the effective CTE $\bar{\alpha}_\theta(\boldsymbol{d})$ is plotted in the x-y plane, obtained by both the FFT-based homogenization and the HS$^+$ method for $k = 1.0$. The corresponding experimentally determined CTE values are represented as red dots.

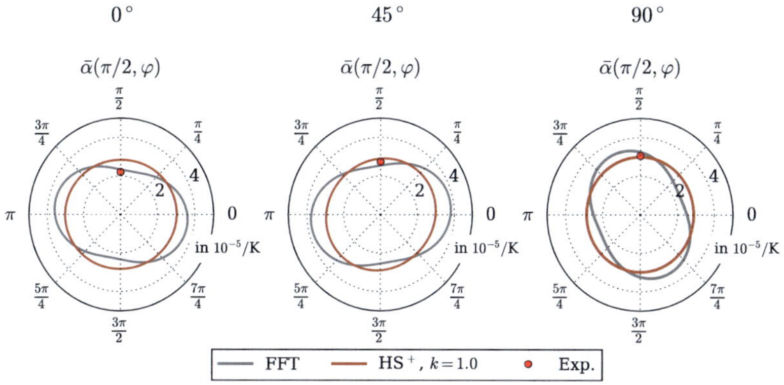

Figure 6.19: Comparison of the effective CTE, obtained by the HS$^+$ method and the FFT method, with the experimental data regarding the 0°, 45°, and 90° samples. The mean-field simulations are performed with $k = 1.0$. The experimental data are depicted for samples from the 1D/1 plaque.

The results of the FFT method exhibit a pronounced ellipsoidal-shaped distribution of $\bar{\alpha}_\theta(\boldsymbol{d})$. The HS$^+$ method for $k = 1.0$ shows a rather planar isotropic distribution. For the 0° sample, the results of the FFT method are closer to the experiments than the results of mean-field methods are. Regarding the 45° and 90° sample, both methods yield results that are in good agreement with the experimental data. The angular distribution of $\bar{\alpha}_\theta(\boldsymbol{d})$ exhibits deviations from the FFT results with respect to the varying polar angle φ. In particular, for the 0° and 45° samples, the mean-field method yields higher $\bar{\alpha}_\theta(\boldsymbol{d})$ values compared to the FFT-based results for φ values near tension direction. The $90°$ sample exhibits an inverse behavior. Analogous to the differences already observed in terms of the predicted effective Young's modulus, this may result from the different input in terms of the orientation of fibers within the composite.

6.5.3 Temperature-dependent effective thermal material properties

The effective CTE for varying temperatures is analyzed analogously to the investigations in Section 6.4. Therefore, a function of $\alpha_{\mathrm{M}}(\theta)$ dependent on the temperature is needed. As discussed in Section 4.2.3, the least square fit using a linear function is reasonable for the measured thermal strains of the matrix material. Thus, a constant value, independent on the temperature, is derived for α_{M}. However, the temperature-dependent elastic properties $E_{\mathrm{M}}(\theta)$ of the matrix are used as input for the computation of the effective thermal behavior. Thus, only Young's modulus of the matrix $E_{\mathrm{M}}(\theta)$ is regarded as temperature-dependent in the following investigations. The material parameters of the fiber $\{E_{\mathrm{F}}, \alpha_{\mathrm{F}}, \nu_{\mathrm{F}}\}$, as well as of the matrix $\{\alpha_{\mathrm{M}}, \nu_{\mathrm{M}}\}$, are assumed to be temperature-independent.

The effective temperature-dependent CTE is computed by means of the HS^+ with $k = 1.0$. The resulting distributions are depicted over the temperature, starting at the reference temperature of $20°\mathrm{C}$ up to $100°\mathrm{C}$. As discussed previously, temperatures above $100°\mathrm{C}$ are out of the range of possible applications and, therefore, not considered here. The mean-field results are compared to the experimental CTE extracted from thermal strain measurements for the composite. In this context, a quadratic least square fit to the measured thermal strain data is considered, leading to a linear distribution of the CTE of the composite material.

Figure 6.20 shows the computed and experimentally obtained CTE for the temperature range considered for the $0°$, $45°$, and $90°$ samples of the plaque 1D/1. The effective CTE exhibits a nearly linear distribution for the different samples. In the temperature ranges of $20°C$ and $50°C$, a flatter slope is observed when compared with the results depicted for temperatures between $50°C$ and $100°C$. The results for the effective stiffness already exhibited deviations from the experimental data, cf. Section 6.4.1. As the effective stiffness is used in the computation of the effective CTE, the deviations observed here are more pronounced. In contrast, the results for the $45°$ and $90°$ samples are close to the experimentally determined distribution. For the $90°$ sample, vanishing deviations are obtained especially for $\theta \geq 50°C$. Thus, using a temperature-dependent $E_{\mathrm{M}}(\theta)$ but assuming a temperature-independent α_{M}, in turn, captures the temperature-dependent effective CTE.

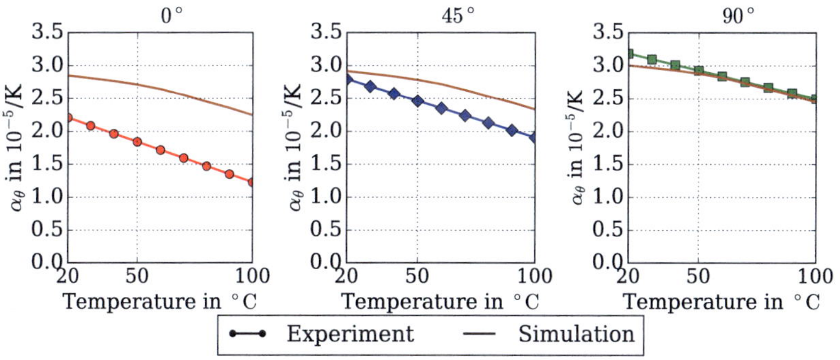

Figure 6.20: Effective temperature-dependent CTE obtained by the HS$^+$ method for $k = 1.0$, and samples of the 1D/1 plaque compared to the experimental CTE over temperature.

6.5.4 Summary of results

To model the thermomechanical material behavior of composites, elastic and thermal material properties of the constituents have to be taken into account. The DMA and CTE measurements provide the basis for computing homogenized solutions for both the effective elastic and effective thermal behavior. In this context, the effective CTE is compared with the experimentally determined CTE. The effective thermal behavior obtained by the HS^+ method at a constant temperature $\theta = 20°C$ are closer to the experimental data than the HS^- results are. The CTE of the fiber material, used in the first step of the HS^+ method, is smaller than the CTE value of the matrix material. Consequently, the effective CTE computed by the HS^+ method is smaller than the results obtained by the HS^- method and, closer to the experimental data. Consistent with the linear elastic investigations, the choice of $k = 1.0$ exhibits the smallest deviation from the experimental CTE. However, to quantify and validate the distribution of $\bar{\alpha}_\theta(\boldsymbol{d})$ for both mean- and full-field homogenization, experimental CTE values at different directions within a sample are needed.

As investigated in Sections 4.1.3 and 4.2.3, the elastic and thermal behavior of the composite exhibits temperature-dependent material parameters. The thermal expansion measurements provide a constant CTE of the matrix material with respect to the temperature. Thus, only Young's modulus of the matrix material is considered temperature-dependent. However, using $E_{\mathrm{M}}(\theta)$ within the computations of the effective CTE, yields a temperature-dependent effective solution. The predicted effective CTEs at the varying temperatures show the highest deviations for the $0°$-oriented sample. A close approximation to the experimental data is obtained for the $45°$ and $90°$ samples.

Chapter 7

Summary and Conclusions

Modern composite materials combine the advantages of a lightweight polymer matrix with those of fiber reinforcements to obtain a high specific stiffness and strength. In the last decades, many composite systems that contain short or long fibers have been developed. A new composite class, which consists of a discontinuous fiber-reinforced thermoset matrix locally reinforced with patches of continuous reinforcements, combines the advantages of both composite systems, and is subject of current research within the research training group DFG-GRK 2078. The thesis at hand specifically addresses discontinuous glass fiber-reinforced thermoset-based composites manufactured of SMCs. The pure and fiber-reinforced thermoset material is experimentally characterized by DMA and thermal expansion measurements. A thermoelastic mean-field model is developed to predict the effective thermal and elastic material behavior. Moreover, the mean-field results are compared to both numerical results obtained by full-field simulations and experimental data. The results of these investigations are summarized in the following.

One of the three main objectives of this thesis is to experimentally characterize the thermoviscoelastic behavior of the SMC material using DMA. This study considers samples of pure matrix material, as well as of reinforced composite material, and investigates their temperature- and frequency-dependent material behavior. The analysis of these tests reveals an anisotropic behavior of the reinforced samples. Both the pure and composite materials show a pronounced temperature dependence of the material properties for the considered temperature range of $-30°C$ to $200°C$. In contrast, the frequency dependence is comparatively less pronounced. An increasing viscoelastic behavior is observed for temperatures exceeding $100°C$. The glass transition temperature is determined, representing the maximum application temperature. In addition, the coefficient of thermal expansion (CTE) is measured in an environmental chamber and using a video extensometer system. Based on the provided database, the CTE of the matrix material is constant throughout the considered temperature range. The CTE of the composite material exhibits a linear behavior.

The experimental data for the matrix material are used as input data for the mean- and full-field simulations to predict the measured thermoelastic behavior of the composite.

The second main objective of this thesis is to develop a completely thermomechanical mean-field method for predicting the effective elastic and thermal properties. Typically, the overall behavior of a fibrous composite is crucially influenced by the heterogeneities on the microscopic level. Hence, the fiber orientation distribution, volume fraction, length distribution, and aspect ratio are important quantities in addition to the elastic and thermal material constants. The SMC composite considered in this thesis consists of long, discontinuous glass fibers at nearly constant lengths of $2.5\,\mathrm{cm}$. The mean fiber orientation is given in terms of orientation tensors of the second- and fourth-order. These tensors are computed directly on the basis of reconstructed voxel-based images obtained by μCT scans (Pinter, 2018). To take the orientation tensors into account, the two-step mean-field method introduced is formulated explicitly in terms of orientation averages. This approach is based on the Hashin-Shtrikman (HS) homogenization method as proposed by Willis (1981) and Walpole (1966). The formulation provided offers a simple and numerically efficient structure. Other mean-field methods, like the Mori-Tanaka, also provide a simple mathematical and computation time-efficient structure. Since non-physical overall results can be obtained for the Mori-Tanka method when applied to anisotropic constituents, it is limited to specific composite classes and thus not suitable for Co or CoDiCo composites. In contrast, the approach developed can be applied to various material systems, and is not limited to isotropic constituents or two-phase composites, for instance.

Within the HS approach, the reference stiffness is commonly chosen to equal either the matrix or the fiber stiffness. However, this yields rather broad bounds due to the high phase contrast between the fiber and the matrix material of the composite considered. Hence, in this thesis, a variable reference stiffness is introduced to obtain a more tailored approach. Moreover, this enables studying the influence of the reference medium on the effective material behavior. In this context, several choices of the reference medium are determined, for which the mean-field approximation is in good agreement with the experimental data previously obtained by DMA and CTE measurements.
Due to the temperature-dependent elastic modulus observed in DMA tests, the HS two-step method is formulated in terms of a temperature-dependent stiffness. Further, measurements provide a constant value for the CTE of the matrix material as input for the homogenization. By applying the HS two-step method, a temperature-dependent effective CTE is obtained due to the temperature-dependent effective stiffness. The elastic and thermal mean-field results are in good agreement with the experimental investigations for reasonable application temperatures.

Additionally, full-field simulations are performed to draw quantitative conclusions on the effective direction-dependent Young's modulus and the CTE obtained by the mean-field simulation. The µCT data are directly imported and segmented using the commercial software tool GeoDict®. Based on these voxel data, the local fields are computed using a fast Fourier transformation. The effective properties for comparing both the mean-field results and the experimental data are determined by volume averaging the local solution. The solutions obtained exhibit vanishing deviations with respect to the experimental elastic and thermal data. Deviations are observed between the solutions obtained by the full-field method and the mean-field results regarding the shape of the orientation-dependent effective Young's modulus, as well as that of the effective thermal expansion. A possible reason for the deviations occurring is given by the different input regarding the fiber orientation for both methods. Moreover, the microstructure characterization reveals clustering effects of the fibers. These clusters, areas of bundled fibers, cannot simply be scanned at a higher resolution to resolve each single fiber filament, since the resulting volume elements are then no longer representative. Thus, a competition between resolution and volume size is given. Prospective studies are advised to generate synthetic volume elements that represent the real microstructure. Based on these structures, a more profound analysis could be performed that allows for a quantitative comparison between the different simulation methods.

The following concluding remarks can be made:

- By means of the dynamic mechanical analysis, the thermoviscoelastic material behavior is characterized. A pronounced temperature-dependent behavior is revealed. Additionally, the coefficient of thermal expansion is identified by thermal expansion measurements of the matrix and the composite material. These experimental findings motivate the development of a thermoelastic mean-field method. Moreover, substantial input, as well as validation data, are provided for simulating the overall thermomechanical behavior.

- The mean-field model presented is derived from the Hashin-Shtrikman scheme, which is enhanced to account for the thermo-mechanical behavior. It provides the basis for considering more complex materials such as carbon fiber-reinforced composites for which methods like the Mori-Tanaka model cannot be generally applied to. Its formulation in terms of orientation tensors of second- and fourth-order and its explicit mathematical structure leads to time-efficient computations. Thus, it is suitable for the integrated use with shape-optimization simulations, for instance, and enables designing structural composite parts. The variable reference stiffness offers the opportunity to adjust the mean-field method to a certain composite class.
- The outcome of the experimental investigations is that the matrix stiffness needs to be modeled temperature-dependent to capture the behavior of the composite over a wide temperature range. The measurements provide a temperature-independent behavior for the CTE of the matrix material. Thus, the temperature-dependency of the effective CTE is induced by the temperature-dependent matrix stiffness, used as input for the mean-field homogenization. In this context, a comprehensive thermomechanical model is developed.
- The FFT-based full-field simulations offer the opportunity to determine the effective elastic and thermal properties directly from µCT data. The deviations observed between the mean- and the full-field results can presumably be traced back to the different orientation input for both methods. Besides, the microstructure is governed by bundled fibers which are poorly resolved and thus affect the full-field results. To overcome this drawback, it is recommended to generate realistic synthetic structures.

For prospective studies, the mean-field method developed could be extended to account for damage and failure mechanisms. Material classes based on thermoplastic matrices exhibit a pronounced viscoelastic behavior. Since the homogenization method presented is based on an eigenstrain formulation, it can be extended to consider the thermoviscoelastic behavior of composites. In the context of long fiber-reinforced thermoplastics, the length distribution of the fibers plays an important role. The length distribution of fibers can be considered within the orientation averaging of this mean-field approach; it can thus easily be applied to long fiber-reinforced thermoplastics.

Appendix A

Mean- and Full-Field Homogenization

A.1 Hill's Polarization Tensor

For an ellipsoidal shaped inclusion as it is illustrated in Figure A.1, Hill's polarization tensor is determined by

$$\mathbb{P}(\mathbb{C}, \boldsymbol{Z}) = \frac{1}{4\pi \det(\boldsymbol{Z})} \int_S \mathbb{H}(\mathbb{C}, \boldsymbol{n})(\boldsymbol{n} \cdot (\boldsymbol{Z}^{-\mathsf{T}} \boldsymbol{Z}^{-1} \boldsymbol{n}))^{-3/2} \, \mathrm{d}S, \tag{A.1}$$

$$\mathbb{H}(\mathbb{C}, \boldsymbol{n}) = \mathbb{I}^{\mathsf{S}}(\boldsymbol{K}^{-1} \square (\boldsymbol{n} \otimes \boldsymbol{n}))\mathbb{I}^{\mathsf{S}}, \qquad \boldsymbol{K} = \mathbb{C}[\![\boldsymbol{n} \otimes \boldsymbol{n}]\!], \tag{A.2}$$

with $S = \{\boldsymbol{n} \in \mathbb{R}^3 : \|\boldsymbol{n}\| = 1\}$. The contraction operator $[\![\cdot]\!]$ is defined by $(\boldsymbol{a} \otimes \boldsymbol{b}) \cdot \mathbb{C}[\![\boldsymbol{a} \otimes \boldsymbol{b}]\!] = (\boldsymbol{a} \otimes \boldsymbol{b}) \cdot \mathbb{C}[\boldsymbol{a} \otimes \boldsymbol{b}]$. The quantity $\boldsymbol{Z}$ describes the ellipsoidal geometry of the fibers. The half axes of the ellipsoid correspond to the reciprocal eigenvalues of $\boldsymbol{Z}$.

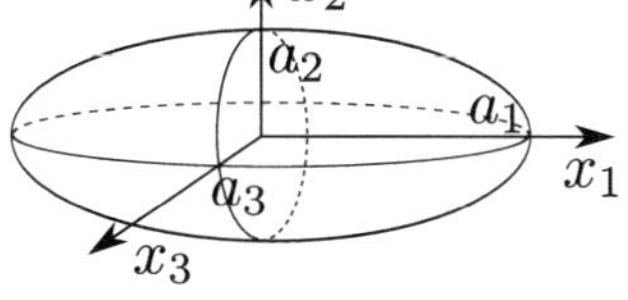

Figure A.1: Ellipsoid with half axes a_1, a_2 and a_3.

Introducing the notation from Walpole (1969), a transversely isotropic tensor with symmetry axis in x_1-direction is represented by

$$\mathbb{P} \widehat{=} (2\alpha, \beta, \gamma, 2\delta, 2\epsilon), \tag{A.3}$$

and thus,

$$\mathbb{P} \widehat{=} \begin{pmatrix} \gamma & \beta & \beta & & & \\ \beta & \alpha+\delta & \alpha-\delta & & & \\ \beta & \alpha-\delta & \alpha+\delta & & & \\ & & & 2\delta & & \\ & & & & 2\epsilon & \\ & & & & & 2\epsilon \end{pmatrix}. \tag{A.4}$$

For the special case of a spherical inclusion, Hill's polarization tensor is given by

$$\mathbb{P}_0 \widehat{=} (2\alpha, \beta, \gamma, 2\delta, 2\epsilon), \tag{A.5}$$

with

$$\begin{aligned} 2\alpha &= \frac{2}{3(3\lambda+6\mu)} + \frac{1}{15}\frac{3\lambda+8\mu}{\mu(3\lambda+6\mu)}, \\ \beta &= \frac{2}{3(3\lambda+6\mu)} - \frac{1}{15}\frac{3\lambda+8\mu}{\mu(3\lambda+6\mu)}, \\ \gamma &= \frac{2}{3(3\lambda+6\mu)} + \frac{2}{15}\frac{3\lambda+8\mu}{\mu(3\lambda+6\mu)}, \\ 2\delta &= \frac{1}{5}\frac{3\lambda+8\mu}{\mu(3\lambda+6\mu)}, \quad 2\epsilon = 2\delta. \end{aligned} \tag{A.6}$$

Furthermore, for the special case of needle shaped, unidirectional inclusions, Hill's polarization tensor reduces to

$$\mathbb{P}^{\mathrm{UD}} \widehat{=} (2\alpha, 0, 0, 2\delta, 2\epsilon), \tag{A.7}$$
$$2\alpha = \frac{1}{2}\frac{1}{\lambda + 2\mu}, \quad 2\delta = \frac{\lambda + 3\mu}{4(\lambda + 2\mu)\mu}, \quad 2\epsilon = \frac{1}{4\mu}.$$

A.2 Hashin-Shtrikman Mean-Field Homogenization

A.2.1 Hashin-Shtrikman functional

The stress-strain relation reads

$$\boldsymbol{\sigma} = \frac{\partial W}{\partial \boldsymbol{\varepsilon}} = \mathbb{C}[\boldsymbol{\varepsilon}] - \boldsymbol{G}, \tag{A.8}$$

with the energy density

$$\begin{aligned} W &= \frac{1}{2}\boldsymbol{\varepsilon} \cdot \mathbb{C}[\boldsymbol{\varepsilon}] - \boldsymbol{\varepsilon} \cdot \boldsymbol{G} - \frac{1}{2}k \\ &= \frac{1}{2}\boldsymbol{\varepsilon} \cdot \boldsymbol{\sigma} - \frac{1}{2}\boldsymbol{\varepsilon} \cdot \boldsymbol{G} - \frac{1}{2}k. \end{aligned} \tag{A.9}$$

The ensemble average is then given by

$$2\langle W \rangle = \langle \boldsymbol{\varepsilon} \cdot \boldsymbol{\sigma} \rangle - \langle \boldsymbol{\varepsilon} \cdot \boldsymbol{G} \rangle - \langle k \rangle. \tag{A.10}$$

According to Willis (1981), a homogeneous comparison medium is introduced with stiffness $\mathbb{C}_0$ for which the polarization field is defined by

$$\boldsymbol{\tau} = (\mathbb{C} - \mathbb{C}_0)[\boldsymbol{\varepsilon}] - \boldsymbol{G} = \delta\mathbb{C}[\boldsymbol{\varepsilon}] - \boldsymbol{G}, \tag{A.11}$$

with $\delta\mathbb{C} = \mathbb{C} - \mathbb{C}_0$. Thus, the stress-strain relation from Eq. (A.8) reads

$$\boldsymbol{\sigma} = \boldsymbol{\tau} + \mathbb{C}_0[\boldsymbol{\varepsilon}]. \tag{A.12}$$

By this, the boundary value problem $\operatorname{div}(\boldsymbol{\sigma}) = \mathbf{0}$ reads

$$\operatorname{div}(\mathbb{C}_0[\boldsymbol{\varepsilon}]) + \operatorname{div}(\boldsymbol{\tau}) = \mathbf{0}. \tag{A.13}$$

Thus, the initial problem without body forces and a heterogeneous material is reformulated to a problem containing homogeneous material and apparent body forces $\boldsymbol{b} = \operatorname{div}(\boldsymbol{\tau})$. The boundary value problem for known polarizations $\boldsymbol{\tau}$ is formally solved by

$$\boldsymbol{\varepsilon} = \bar{\boldsymbol{\varepsilon}} - \mathbb{G}\{\boldsymbol{\tau}\}, \tag{A.14}$$

with $\mathbb{G}$ denoting the integral Green operator, cf. Willis (1981; 1977), defined by $\mathbb{G}\{\boldsymbol{\tau}\} = \int_\Omega \mathbb{G}(\boldsymbol{x}, \boldsymbol{x}')\boldsymbol{\tau}(\boldsymbol{x}')\,\mathrm{d}v'$. Combining Eqs. (A.11) and (A.14), yields

$$\delta\mathbb{C}^{-1}(\boldsymbol{\tau} + \boldsymbol{G}) = \bar{\boldsymbol{\varepsilon}} - \mathbb{G}\{\boldsymbol{\tau}\} \tag{A.15}$$

$$\bar{\boldsymbol{\varepsilon}} - \delta\mathbb{C}^{-1}[\boldsymbol{G}] = \delta\mathbb{C}^{-1}[\boldsymbol{\tau}] + \mathbb{G}\{\boldsymbol{\tau}\} \tag{A.16}$$

$$\tilde{\boldsymbol{\varepsilon}} = \delta\mathbb{C}^{-1}[\boldsymbol{\tau}] + \mathbb{G}\{\boldsymbol{\tau}\}, \tag{A.17}$$

with $\tilde{\boldsymbol{\varepsilon}} = \bar{\boldsymbol{\varepsilon}} - \delta\mathbb{C}^{-1}[\boldsymbol{G}]$. Rearranging yields the consistency condition reading

$$\boldsymbol{K} := \delta\mathbb{C}^{-1}[\boldsymbol{\tau}] + \mathbb{G}\{\boldsymbol{\tau}\} - \tilde{\boldsymbol{\varepsilon}} = \mathbf{0}, \tag{A.18}$$

for which $\boldsymbol{K}(\boldsymbol{\tau} = \hat{\boldsymbol{\tau}}) = \mathbf{0}$ for exact polarizations $\hat{\boldsymbol{\tau}}$. Using $\boldsymbol{K}$, the variation functional according to Hashin and Shtrikman (1962) considering eigenfields is given by

$$\mathcal{F} = \langle \frac{1}{2}\boldsymbol{\tau} \cdot \delta\mathbb{C}^{-1}[\boldsymbol{\tau}] + \frac{1}{2}\boldsymbol{\tau} \cdot \mathbb{G}\{\boldsymbol{\tau}\} - \boldsymbol{\tau} \cdot \tilde{\boldsymbol{\varepsilon}} \rangle \tag{A.19}$$

$$= \frac{1}{2}\langle \boldsymbol{\tau} \cdot \boldsymbol{K} \rangle - \frac{1}{2}\langle \boldsymbol{\tau} \cdot \tilde{\boldsymbol{\varepsilon}} \rangle. \tag{A.20}$$

The first variation reads

$$\delta\mathcal{F} = \langle \boldsymbol{K} \cdot \delta\boldsymbol{\tau} \rangle, \tag{A.21}$$

which is stationary for exact polarizations $\hat{\boldsymbol{\tau}}$. By means of the second variation, the minimum and maximum of the HS functional $\mathcal{F}$ is determined by

$$\delta^2\mathcal{F} = \langle \delta\boldsymbol{\tau} \cdot (\delta\mathbb{C}^{-1} + \mathbb{G})[\boldsymbol{\tau}] \rangle, \tag{A.22}$$

for which

$$\delta^2\mathcal{F} = \begin{cases} > 0, & \mathcal{F}\,\text{min}, \\ < 0, & \mathcal{F}\,\text{max}, \end{cases} \tag{A.23}$$

is obtained. According to Willis (1977), a positive (negative) definite $\delta\mathbb{C}$ implies a positive (negative) definite $\delta\mathbb{C}^{-1} + \mathbb{G}$. For an isotropic homogeneous comparison material with stiffness $\mathbb{C}_0$, the Green operator simplifies to $\mathbb{G}\{\boldsymbol{\tau}\} \approx \mathbb{P}_0[\boldsymbol{\tau}]$, with $\mathbb{P}_0$ denoting Hill's polarization tensor for an isotropic medium.

The value of $\mathcal{F}$ for exact polarizations $\hat{\boldsymbol{\tau}}$ reads

$$\begin{aligned}
\mathcal{F}(\hat{\boldsymbol{\tau}}) &= -\frac{1}{2}\langle \hat{\boldsymbol{\tau}} \cdot \tilde{\varepsilon} \rangle && \text{(A.24)}\\
&= -\frac{1}{2}\bar{\boldsymbol{\sigma}} \cdot \bar{\varepsilon} + \frac{1}{2}\mathbb{C}_0[\bar{\varepsilon}] \cdot \bar{\varepsilon} + \frac{1}{2}\langle \boldsymbol{G} \cdot \varepsilon \rangle - \frac{1}{2}\langle \boldsymbol{G} \cdot \delta\mathbb{C}^{-1}[\boldsymbol{G}] \rangle && \text{(A.25)}\\
&\qquad - \frac{1}{2}\langle k \rangle + \frac{1}{2}\langle k \rangle \\
&= -\bar{W} + \frac{1}{2}\mathbb{C}_0[\bar{\varepsilon}] \cdot \bar{\varepsilon} - \frac{1}{2}\langle \boldsymbol{G} \cdot \delta\mathbb{C}^{-1}[\boldsymbol{G}] \rangle - \frac{1}{2}\langle k \rangle && \text{(A.26)}\\
&= -\bar{W} + \bar{W}_0. && \text{(A.27)}
\end{aligned}$$

Thus, from the HS principle, it follows

$$\bar{W}^{\text{HS}+} = \bar{W}_0 - \max_{\boldsymbol{\tau}} \mathcal{F}(\boldsymbol{\tau}) \geq \bar{W} \qquad \text{(A.28)}$$

$$\bar{W}^{\text{HS}-} = \bar{W}_0 - \min_{\boldsymbol{\tau}} \mathcal{F}(\boldsymbol{\tau}) \leq \bar{W}. \qquad \text{(A.29)}$$

Formally,

$$\bar{W}^{\text{HS}} = \frac{1}{2}\bar{\varepsilon} \cdot \bar{\mathbb{C}}^{\text{HS}}[\bar{\varepsilon}] - \bar{\varepsilon} \cdot \bar{\boldsymbol{G}}^{\text{HS}} - \frac{1}{2}\bar{k}^{\text{HS}}, \qquad \text{(A.30)}$$

and furthermore with

$$\begin{aligned}
\bar{W}^{\text{HS}} &= \bar{W}_0 - \mathcal{F}(\boldsymbol{\tau}) = \bar{W}_0 + \frac{1}{2}\langle \boldsymbol{\tau} \cdot \tilde{\varepsilon} \rangle && \text{(A.31)}\\
&= \frac{1}{2}\mathbb{C}_0[\bar{\varepsilon}] \cdot \bar{\varepsilon} - \frac{1}{2}\langle \boldsymbol{G} \cdot \delta\mathbb{C}^{-1}[\boldsymbol{G}] \rangle - \frac{1}{2}\langle k \rangle + \frac{1}{2}\boldsymbol{\tau} \cdot \tilde{\varepsilon} && \text{(A.32)}
\end{aligned}$$

and $\varepsilon = \mathbb{A}[\bar{\varepsilon}] - \boldsymbol{a}$, a comparison of coefficients yields

$$\bar{\mathbb{C}}^{\text{HS}} = \langle \mathbb{C}_0 + \delta\mathbb{C}\mathbb{A} \rangle = \langle \mathbb{C}\mathbb{A} \rangle \qquad \text{(A.33)}$$

$$\bar{\boldsymbol{G}}^{\text{HS}} = \langle \mathbb{A}^{\mathsf{T_H}}[\boldsymbol{G}] \rangle. \qquad \text{(A.34)}$$

From Eq. (A.33), in the first step for a domain γ with unidirectional polarization tensor $\mathbb{P}^{\mathrm{UD}}$, $\mathbb{C}_0 = \mathbb{C}_{\mathrm{M}}$ is chosen leading to $\bar{W}^{\mathrm{HS}} = \bar{W}^{\mathrm{HS}-}$ and thus, the effective stiffness is given by

$$\begin{aligned}\bar{\mathbb{C}}_{\gamma}^{\mathrm{UD}-} &= \mathbb{C}_{\mathrm{M}} + \langle \delta\mathbb{C}\mathbb{A}\rangle \\ &= \mathbb{C}_{\mathrm{M}} + c_{\gamma}(\mathbb{C}_{\gamma} - \mathbb{C}_{\mathrm{M}})\mathbb{A}_{\gamma}^{\mathrm{HS}}. \end{aligned} \tag{A.35}$$

With $\mathbb{A}_{\gamma}^{\mathrm{HS}} = \mathbb{M}_{\gamma}\langle\mathbb{M}\rangle^{-1}$, $\mathbb{M}_{\gamma} = (\mathbb{I}^{\mathsf{S}} + \mathbb{P}_{\gamma}^{\mathrm{UD}}\delta\mathbb{C}_{\gamma})^{-1}$ and $\delta\mathbb{C}_{\gamma} = \mathbb{C}_{\gamma} - \mathbb{C}_0 = \mathbb{C}_{\gamma} - \mathbb{C}_{\mathrm{M}}$, it follows

$$\begin{aligned}\bar{\mathbb{C}}_{\gamma}^{\mathrm{UD}-} &= \mathbb{C}_{\mathrm{M}} + c_{\gamma}\delta\mathbb{C}_{\gamma}(\mathbb{I}^{\mathsf{S}} + \mathbb{P}_{\gamma}^{\mathrm{UD}}\delta\mathbb{C}_{\gamma})^{-1}\langle(\mathbb{I}^{\mathsf{S}} + \mathbb{P}_{\gamma}^{\mathrm{UD}}\delta\mathbb{C}_{\gamma})^{-1}\rangle^{-1} \\ &= \mathbb{C}_{\mathrm{M}} + c_{\gamma}\delta\mathbb{C}_{\gamma}(\mathbb{I}^{\mathsf{S}} + \mathbb{P}_{\gamma}^{\mathrm{UD}}\delta\mathbb{C}_{\gamma})^{-1}(c_{\mathrm{M}}(\mathbb{I}^{\mathsf{S}} + \mathbb{P}_{\gamma}^{\mathrm{UD}}(\mathbb{C}_{\mathrm{M}} - \mathbb{C}_{\mathrm{M}}))^{-1} \\ &\qquad + c_{\gamma}(\mathbb{I}^{\mathsf{S}} + \mathbb{P}_{\gamma}^{\mathrm{UD}}\delta\mathbb{C}_{\gamma})^{-1})^{-1} \\ &= \mathbb{C}_{\mathrm{M}} + c_{\gamma}\delta\mathbb{C}_{\gamma}(\mathbb{I}^{\mathsf{S}} + \mathbb{P}_{\gamma}^{\mathrm{UD}}\delta\mathbb{C}_{\gamma})^{-1}(c_{\mathrm{M}}\mathbb{I}^{\mathsf{S}} + c_{\gamma}(\mathbb{I}^{\mathsf{S}} + \mathbb{P}^{\mathrm{UD}}\delta\mathbb{C}_{\gamma})^{-1})^{-1} \\ &= \mathbb{C}_{\mathrm{M}} + c_{\gamma}\delta\mathbb{C}_{\gamma}((c_{\mathrm{M}}\mathbb{I}^{\mathsf{S}} + c_{\gamma}(\mathbb{I}^{\mathsf{S}} + \mathbb{P}^{\mathrm{UD}}\delta\mathbb{C}_{\gamma})^{-1})(\mathbb{I}^{\mathsf{S}} + \mathbb{P}^{\mathrm{UD}}\delta\mathbb{C}_{\gamma}))^{-1} \\ &= \mathbb{C}_{\mathrm{M}} + c_{\gamma}\delta\mathbb{C}_{\gamma}(c_{\mathrm{M}}(\mathbb{I}^{\mathsf{S}} + \mathbb{P}^{\mathrm{UD}}\delta\mathbb{C}_{\gamma}) + c_{\gamma}\mathbb{I}^{\mathsf{S}})^{-1} \\ &= \mathbb{C}_{\mathrm{M}} + c_{\gamma}\delta\mathbb{C}_{\gamma}((c_{\mathrm{M}} + c_{\gamma})\mathbb{I}^{\mathsf{S}} + c_{\mathrm{M}}\mathbb{P}^{\mathrm{UD}}\delta\mathbb{C}_{\gamma})^{-1} \\ &= \mathbb{C}_{\mathrm{M}} + c_{\gamma}\delta\mathbb{C}_{\gamma}(\mathbb{I}^{\mathsf{S}} + c_{\mathrm{M}}\mathbb{P}^{\mathrm{UD}}\delta\mathbb{C}_{\gamma})^{-1}. \end{aligned} \tag{A.36}$$

Analogous calculations with $\mathbb{C}_0 = \mathbb{C}_{\gamma}$ within the first step yield $\bar{\mathbb{C}}^{\mathrm{UD}+}$. The effective eigenstress contribution is determined by

$$\begin{aligned}\bar{\boldsymbol{G}}^{\mathrm{HS}} &= \langle \mathbb{A}^{\mathsf{T_H}}[\boldsymbol{G}]\rangle \\ &= \langle \mathbb{A}^{\mathsf{T_H}}\mathbb{C}[\boldsymbol{\varepsilon}^*]\rangle \\ \bar{\mathbb{C}}^{\mathrm{HS}}[\bar{\boldsymbol{\alpha}}^{\mathrm{HS}}] &= \langle \mathbb{A}^{\mathsf{T_H}}\mathbb{C}[\boldsymbol{\alpha}]\rangle \\ \bar{\boldsymbol{\alpha}}^{\mathrm{HS}} &= (\bar{\mathbb{C}}^{\mathrm{HS}})^{-1}\langle \mathbb{A}^{\mathsf{T_H}}\mathbb{C}[\boldsymbol{\alpha}]\rangle. \end{aligned} \tag{A.37}$$

For the domain γ and choosing $\mathbb{C}_0 = \mathbb{C}_{\mathrm{M}}$, the effective CTE is determined in the first step by

$$\begin{aligned}
\bar{\boldsymbol{\alpha}}_{\gamma}^{\mathrm{UD}-} &= (\bar{\mathbb{C}}^{\mathrm{UD}-})^{-1}(\mathbb{C}_{\mathrm{M}}[\boldsymbol{\alpha}_{\mathrm{M}}] + c_{\gamma}\mathbb{A}_{\gamma}^{\mathsf{T_H}}(\mathbb{C}_{\gamma}[\boldsymbol{\alpha}_{\gamma}] - \mathbb{C}_{\mathrm{M}}[\boldsymbol{\alpha}_{\mathrm{M}}])) \\
&= (\bar{\mathbb{C}}^{\mathrm{UD}-})^{-1}(\mathbb{C}_{\mathrm{M}}[\boldsymbol{\alpha}_{\mathrm{M}}] + c_{\gamma}\langle(\mathbb{I}^{\mathsf{S}} + \mathbb{P}^{\mathrm{UD}}\delta\mathbb{C}_{\gamma})^{-1}\rangle^{-\mathsf{T_H}} \\
&\qquad (\mathbb{I}^{\mathsf{S}} + \mathbb{P}^{\mathrm{UD}}\delta\mathbb{C}_{\gamma})^{-\mathsf{T_H}}(\mathbb{C}_{\gamma}[\boldsymbol{\alpha}_{\gamma}] - \mathbb{C}_{\mathrm{M}}[\boldsymbol{\alpha}_{\mathrm{M}}])) \\
&= (\bar{\mathbb{C}}^{\mathrm{UD}-})^{-1}(\mathbb{C}_{\mathrm{M}}[\boldsymbol{\alpha}_{\mathrm{M}}] + c_{\gamma}(c_{\mathrm{M}}\mathbb{I}^{\mathsf{S}} + c_{\gamma}(\mathbb{I}^{\mathsf{S}} + \mathbb{P}^{\mathrm{UD}}\delta\mathbb{C}_{\gamma})^{-1})^{-\mathsf{T_H}} \\
&\qquad (\mathbb{I}^{\mathsf{S}} + \mathbb{P}^{\mathrm{UD}}\delta\mathbb{C}_{\gamma})^{-\mathsf{T_H}}(\mathbb{C}_{\gamma}[\boldsymbol{\alpha}_{\gamma}] - \mathbb{C}_{\mathrm{M}}[\boldsymbol{\alpha}_{\mathrm{M}}])) \\
&= (\bar{\mathbb{C}}^{\mathrm{UD}-})^{-1}(\mathbb{C}_{\mathrm{M}}[\boldsymbol{\alpha}_{\mathrm{M}}] + c_{\gamma}((c_{\mathrm{M}}\mathbb{I}^{\mathsf{S}} + c_{\gamma}(\mathbb{I}^{\mathsf{S}} + \mathbb{P}^{\mathrm{UD}}\delta\mathbb{C}_{\gamma})^{-1}) \\
&\qquad (\mathbb{I}^{\mathsf{S}} + \mathbb{P}^{\mathrm{UD}}\delta\mathbb{C}_{\gamma}))^{-\mathsf{T_H}}(\mathbb{C}_{\gamma}[\boldsymbol{\alpha}_{\gamma}] - \mathbb{C}_{\mathrm{M}}[\boldsymbol{\alpha}_{\mathrm{M}}])) \\
&= (\bar{\mathbb{C}}^{\mathrm{UD}-})^{-1}(\mathbb{C}_{\mathrm{M}}[\boldsymbol{\alpha}_{\mathrm{M}}] + c_{\gamma}((c_{\mathrm{M}} + c_{\gamma})\mathbb{I}^{\mathsf{S}} + c_{\mathrm{M}}\mathbb{P}^{\mathrm{UD}}\delta\mathbb{C}_{\gamma})^{-\mathsf{T_H}} \\
&\qquad (\mathbb{C}_{\gamma}[\boldsymbol{\alpha}_{\gamma}] - \mathbb{C}_{\mathrm{M}}[\boldsymbol{\alpha}_{\mathrm{M}}])) \\
&= (\bar{\mathbb{C}}^{\mathrm{UD}-})^{-1}(\mathbb{C}_{\mathrm{M}}[\boldsymbol{\alpha}_{\mathrm{M}}] + c_{\gamma}(\mathbb{I}^{\mathsf{S}} + c_{\mathrm{M}}\mathbb{P}^{\mathrm{UD}}\delta\mathbb{C}_{\gamma})^{-\mathsf{T_H}} \\
&\qquad (\mathbb{C}_{\gamma}[\boldsymbol{\alpha}_{\gamma}] - \mathbb{C}_{\mathrm{M}}[\boldsymbol{\alpha}_{\mathrm{M}}]))
\end{aligned} \tag{A.38}$$

Analogous calculations with $\mathbb{C}_0 = \mathbb{C}_{\gamma}$ within the first step yields $\bar{\boldsymbol{\alpha}}^{\mathrm{UD}+}$.

The effective stiffness is calculated by

$$\begin{aligned}
\bar{\mathbb{C}} &= \langle \mathbb{C}\mathbb{A}^{\mathrm{HS}} \rangle \\
&= \langle \mathbb{C}\mathbb{M} \rangle \langle \mathbb{M} \rangle^{-1} \\
&= \langle \delta\mathbb{C}\mathbb{M} \rangle \langle \mathbb{M} \rangle^{-1} + \langle \mathbb{C}_0\mathbb{M} \rangle \langle \mathbb{M} \rangle^{-1} \\
&= \langle \delta\mathbb{C}\mathbb{M} \rangle \langle \mathbb{M} \rangle^{-1} + \mathbb{C}_0 \langle \mathbb{M} \rangle \langle \mathbb{M} \rangle^{-1} \\
&= \mathbb{C}_0 + \langle \delta\mathbb{C}\mathbb{M} \rangle \langle \mathbb{M} \rangle^{-1} \\
&= \mathbb{C}_0 + \langle \mathbb{P}_0^{-1}\mathbb{P}_0\delta\mathbb{C}\mathbb{M} \rangle \langle \mathbb{M} \rangle^{-1} \\
&= \mathbb{C}_0 + \langle \mathbb{P}_0^{-1}(-\mathbb{I}^{\mathsf{S}} + \mathbb{I}^{\mathsf{S}} + \mathbb{P}_0\delta\mathbb{C})\mathbb{M} \rangle \langle \mathbb{M} \rangle^{-1} \\
&= \mathbb{C}_0 + \langle \mathbb{P}_0^{-1}(-\mathbb{I}^{\mathsf{S}} + \mathbb{M}^{-1})\mathbb{M} \rangle \langle \mathbb{M} \rangle^{-1} \\
&= \mathbb{C}_0 + \langle \mathbb{P}_0^{-1} - \mathbb{P}_0^{-1}\mathbb{M} \rangle \langle \mathbb{M} \rangle^{-1} \\
&= \mathbb{C}_0 + \mathbb{P}_0^{-1} \langle \mathbb{M} \rangle^{-1} - \mathbb{P}_0^{-1} \\
&= \mathbb{C}_0 - \mathbb{P}_0^{-1} + \langle \mathbb{M}\mathbb{P}_0 \rangle^{-1} \\
&= \mathbb{C}_0 - \mathbb{P}_0^{-1} + \langle (\mathbb{I}^{\mathsf{S}} + \mathbb{P}_0\delta)^{-1}\mathbb{P}_0 \rangle^{-1} \\
&= \mathbb{C}_0 - \mathbb{P}_0^{-1} + \langle (\mathbb{P}_0^{-1} + \delta\mathbb{C})^{-1} \rangle^{-1}.
\end{aligned} \tag{A.39}$$

The latter equation is reformulated according to Walpole (1966; 1969) to obtain a computational efficient and simple expression for the effective stiffness. The effective stiffness within the second step of the HS method is determined by

$$\begin{aligned}
\bar{\mathbb{C}}^{\mathrm{HS}\pm} &= \langle \frac{1}{c_{\mathrm{F}}} \bar{\mathbb{C}}^{\mathrm{UD}\pm} \mathbb{A}^{\mathrm{HS}} \rangle \\
&= \frac{1}{c_{\mathrm{F}}} \sum_{\beta=1}^{N} c_\beta \mathbb{C}_\beta^{\mathrm{UD}\pm} \mathbb{A}_\beta \\
&= \mathbb{C}_0 - \mathbb{P}_0^{-1} + \langle \frac{1}{c_{\mathrm{F}}} (\mathbb{P}_0^{-1} + \bar{\mathbb{C}}^{\mathrm{UD}\pm} - \mathbb{C}_0)^{-1} \rangle^{-1}
\end{aligned} \tag{A.40}$$

A.2.2 Effective quantities

The effective energy density reads

$$
\begin{aligned}
\bar{W} &= \frac{1}{2}\langle \varepsilon \cdot \boldsymbol{\sigma} \rangle - \frac{1}{2}\langle \varepsilon \cdot \boldsymbol{G} \rangle - \frac{1}{2}\langle k \rangle \\
&= \frac{1}{2}\langle \varepsilon \cdot \mathbb{C}[\varepsilon] \rangle - \langle \boldsymbol{G} \cdot \varepsilon \rangle - \frac{1}{2}\langle k \rangle \\
&= \frac{1}{2}\langle \mathbb{A}[\bar{\varepsilon}] \cdot \mathbb{C}\mathbb{A}[\bar{\varepsilon}] - \mathbb{A}[\bar{\varepsilon}] \cdot \mathbb{C}[\boldsymbol{a}] - \boldsymbol{a} \cdot \mathbb{C}\mathbb{A}[\bar{\varepsilon}] + \boldsymbol{a} \cdot \mathbb{C}[\boldsymbol{a}] \rangle \\
&\qquad\qquad - \langle \boldsymbol{G} \cdot \mathbb{A}[\bar{\varepsilon}] - \boldsymbol{G} \cdot \boldsymbol{a} \rangle - \frac{1}{2}\langle k \rangle \\
&= \frac{1}{2}\langle \bar{\varepsilon} \cdot \mathbb{A}^{\mathsf{T_H}}\mathbb{C}\mathbb{A}[\bar{\varepsilon}] \rangle - \frac{1}{2}\langle \boldsymbol{a} \cdot \mathbb{C}\mathbb{A}[\bar{\varepsilon}] \rangle - \frac{1}{2}\langle \boldsymbol{a} \cdot \mathbb{C}\mathbb{A}[\bar{\varepsilon}] \rangle + \\
&\qquad\qquad \frac{1}{2}\langle \boldsymbol{a} \cdot \mathbb{C}[\boldsymbol{a}] \rangle - \langle \mathbb{A}^{\mathsf{T_H}}[\boldsymbol{G}] \cdot \bar{\varepsilon} \rangle + \langle \boldsymbol{G} \cdot \boldsymbol{a} \rangle - \frac{1}{2}\langle k \rangle,
\end{aligned}
\tag{A.41}
$$

with $\varepsilon = \mathbb{A}[\bar{\varepsilon}] - \boldsymbol{a}$.
Considering the Hill-Mandel relations, $\langle \mathbb{C}\mathbb{A}[\bar{\varepsilon}] \cdot \boldsymbol{a} \rangle = \langle \mathbb{C}\mathbb{A} \rangle[\bar{\varepsilon}] \cdot \langle \boldsymbol{a} \rangle = \boldsymbol{0}$ and $\langle (\mathbb{C}[\boldsymbol{a}] + \boldsymbol{G}) \cdot \boldsymbol{a} \rangle = \langle \mathbb{C}[\boldsymbol{a}] + \boldsymbol{G} \rangle \cdot \langle \boldsymbol{a} \rangle = \boldsymbol{0}$, the effective density yields

$$
\bar{W} = \frac{1}{2}\langle \bar{\varepsilon} \cdot \mathbb{A}^{\mathsf{T_H}}\mathbb{C}\mathbb{A}[\bar{\varepsilon}] \rangle - \langle \mathbb{A}^{\mathsf{T_H}}[\boldsymbol{G}] \rangle \cdot \bar{\varepsilon} - \frac{1}{2}\langle k \rangle. \tag{A.42}
$$

With $\bar{W} = \bar{\varepsilon} \cdot \bar{\mathbb{C}}[\bar{\varepsilon}]/2 - \bar{\boldsymbol{G}} \cdot \bar{\varepsilon} - \bar{k}/2$, one yields from Eq. (A.42)

$$
\bar{\mathbb{C}} = \langle \mathbb{A}^{\mathsf{T_H}}\mathbb{C}\mathbb{A} \rangle, \qquad \bar{\boldsymbol{G}} = \langle \mathbb{A}^{\mathsf{T_H}}[\boldsymbol{G}] \rangle \tag{A.43}
$$

A.2.3 Mechanical strain and eigenstrain localization

Given is Hooke's law including eigenstresses $\boldsymbol{G}$

$$
\boldsymbol{\sigma} = \mathbb{C}[\varepsilon] - \boldsymbol{G}. \tag{A.44}
$$

By $\operatorname{div}(\boldsymbol{\sigma}) = \mathbf{0}$ and $\boldsymbol{\tau} = \delta\mathbb{C}[\boldsymbol{\varepsilon}] - \delta\boldsymbol{G}$, an equivalent boundary value problem is derived for which the solution formally reads

$$\begin{aligned}\boldsymbol{\varepsilon} &= \bar{\boldsymbol{\varepsilon}} - \mathbb{G}\{\boldsymbol{\tau}\} \\ &= \bar{\boldsymbol{\varepsilon}} - \mathbb{G}\{\boldsymbol{\tau} - \langle\boldsymbol{\tau}\rangle\}. \end{aligned} \tag{A.45}$$

For an isotropic material, $\mathbb{G}\{\boldsymbol{\tau}\} \approx \mathbb{P}_0[\boldsymbol{\tau}]$ holds, with $\mathbb{P}_0$ denoting Hill's polarization tensor for an isotropic comparison material with stiffness $\mathbb{C}_0$. Thus, one yields

$$\begin{aligned}\delta\mathbb{C}^{-1}[\boldsymbol{\tau}] + \mathbb{C}^{-1}[\delta\boldsymbol{G}] &= \bar{\boldsymbol{\varepsilon}} - \mathbb{P}_0[\boldsymbol{\tau} - \langle\boldsymbol{\tau}\rangle] \\ (\delta\mathbb{C}^{-1} + \mathbb{P}_0)[\boldsymbol{\tau}] - \mathbb{P}_0[\langle\boldsymbol{\tau}\rangle] &= \bar{\boldsymbol{\varepsilon}} - \delta\mathbb{C}^{-1}[\delta\boldsymbol{G}] \\ (\delta\mathbb{C}^{-1} + \mathbb{P}_0)(\delta\mathbb{C}[\boldsymbol{\varepsilon}] - \delta\boldsymbol{G}) - \mathbb{P}_0\langle\delta\mathbb{C}[\boldsymbol{\varepsilon}] - \delta\boldsymbol{G}\rangle &= \bar{\boldsymbol{\varepsilon}} - \delta\mathbb{C}^{-1}[\delta\boldsymbol{G}] \\ (\mathbb{I}^{\mathsf{S}} + \mathbb{P}_0\delta\mathbb{C})[\boldsymbol{\varepsilon}] - \delta\mathbb{C}^{-1}[\delta\boldsymbol{G}] - \mathbb{P}_0[\delta\boldsymbol{G}] - \mathbb{P}_0\langle\delta\mathbb{C}[\boldsymbol{\varepsilon}] - \delta\boldsymbol{G}\rangle &= \bar{\boldsymbol{\varepsilon}} - \delta\mathbb{C}^{-1}[\delta\boldsymbol{G}] \\ (\mathbb{I}^{\mathsf{S}} + \mathbb{P}_0\delta\mathbb{C})[\boldsymbol{\varepsilon}] - \langle(\mathbb{I}^{\mathsf{S}} + \mathbb{P}_0\delta\mathbb{C})[\boldsymbol{\varepsilon}]\rangle &= \mathbb{P}_0[\delta\boldsymbol{G} - \langle\delta\boldsymbol{G}\rangle]. \end{aligned} \tag{A.46}$$

Introducing $\mathbb{M} = (\mathbb{I}^{\mathsf{S}} + \mathbb{P}_0\delta\mathbb{C})^{-1}$ and $\mathbb{D} = \mathbb{P}_0[\delta\boldsymbol{G}]$, the expressions above yield

$$\mathbb{M}^{-1}[\boldsymbol{\varepsilon}] - \langle\mathbb{M}^{-1}[\boldsymbol{\varepsilon}]\rangle = \mathbb{D} - \langle\mathbb{D}\rangle. \tag{A.47}$$

The solution is given by

$$\boldsymbol{\varepsilon} = \mathbb{A}[\bar{\boldsymbol{\varepsilon}}] - \boldsymbol{a}, \qquad \mathbb{A} = \mathbb{M}\langle\mathbb{M}\rangle^{-1}, \qquad \mathbb{M} = (\mathbb{I}^{\mathsf{S}} + \mathbb{P}_0\delta\mathbb{C})^{-1}. \tag{A.48}$$

Combining Eq. (A.48) and Eq. (A.47) yields

$$\begin{aligned}\mathbb{M}^{-1}(\mathbb{M}\langle\mathbb{M}\rangle^{-1}[\bar{\boldsymbol{\varepsilon}}] - \boldsymbol{a}) - \langle\mathbb{M}^{-1}(\mathbb{M}\langle\mathbb{M}\rangle^{-1}[\bar{\boldsymbol{\varepsilon}}] - \boldsymbol{a})\rangle &= \mathbb{D} - \langle\mathbb{D}\rangle \\ \langle\mathbb{M}\rangle^{-1}[\bar{\boldsymbol{\varepsilon}}] - \mathbb{M}^{-1}[\boldsymbol{a}] - \langle\langle\mathbb{M}\rangle^{-1}[\bar{\boldsymbol{\varepsilon}}] - \mathbb{M}^{-1}[\boldsymbol{a}]\rangle &= \mathbb{D} - \langle\mathbb{D}\rangle \\ \langle\mathbb{M}^{-1}[\boldsymbol{a}]\rangle - \mathbb{M}^{-1}[\boldsymbol{a}] &= \mathbb{D} - \langle\mathbb{D}\rangle. \end{aligned} \tag{A.49}$$

In order to solve the latter equation, $\boldsymbol{a} = -\mathbb{M}\mathbb{D}$ is assumed, leading to

$$\begin{aligned}\langle -\mathbb{M}^{-1}\mathbb{M}\mathbb{D}\rangle + \mathbb{M}^{-1}\mathbb{M}\mathbb{D} &= \mathbb{D} - \langle \mathbb{D}\rangle \\ -\langle \mathbb{D}\rangle + \mathbb{D} &= \mathbb{D} - \langle \mathbb{D}\rangle,\end{aligned} \tag{A.50}$$

resulting to a true statement.

From Eq. (A.48)$_{2,3}$, it follows

$$\begin{aligned}\mathbb{A} &= \mathbb{M}\mathbb{P}_0\mathbb{P}_0^{-1}\langle \mathbb{M}\rangle^{-1} \\ &= (\mathbb{I}^{\mathsf{S}} + \mathbb{P}_0\delta\mathbb{C})^{-1}\mathbb{P}_0\mathbb{P}_0^{-1}\langle(\mathbb{I}^{\mathsf{S}} + \mathbb{P}_0\delta\mathbb{C})^{-1}\rangle^{-1} \\ &= (\mathbb{P}_0^{-1} + \delta\mathbb{C})^{-1}\langle(\mathbb{P}_0^{-1} + \delta\mathbb{C})^{-1}\rangle^{-1} \\ &= \mathbb{R}\langle \mathbb{R}\rangle^{-1},\end{aligned} \tag{A.51}$$

with $\mathbb{R} = (\mathbb{P}_0^{-1} + \delta\mathbb{C})^{-1}$. Thus, for the eigenstrain localization it follows

$$\begin{aligned}\boldsymbol{a} &= -\mathbb{M}\mathbb{D} \\ &= -\mathbb{M}\mathbb{P}_0[\delta\boldsymbol{G}] \\ &= -\mathbb{M}\mathbb{P}_0[\boldsymbol{G} - \boldsymbol{G}_0] \\ &= -\mathbb{M}\mathbb{P}_0[\boldsymbol{G}] + \mathbb{M}\mathbb{P}_0[\boldsymbol{G}_0].\end{aligned} \tag{A.52}$$

Using the normalization condition $\langle \boldsymbol{a}\rangle = \boldsymbol{0}$, the reference eigenstrain field $\boldsymbol{G}_0$ can be eliminated, yielding

$$\begin{aligned}\boldsymbol{0} &= \langle -\mathbb{M}\mathbb{P}_0[\boldsymbol{G}]\rangle + \langle \mathbb{M}\rangle\mathbb{P}_0[\boldsymbol{G}_0] \\ \langle \mathbb{M}\rangle\mathbb{P}_0[\boldsymbol{G}_0] &= \langle \mathbb{M}\mathbb{P}_0[\boldsymbol{G}]\rangle \\ \boldsymbol{G}_0 &= \mathbb{P}_0^{-1}\langle \mathbb{M}\rangle^{-1}\langle \mathbb{M}\mathbb{P}_0[\boldsymbol{G}]\rangle\end{aligned} \tag{A.53}$$

Inserting in Eq. (A.52), one obtains

$$\begin{aligned} \boldsymbol{a} &= -\mathbb{M}\mathbb{P}_0[\boldsymbol{G}] + \mathbb{M}\mathbb{P}_0\mathbb{P}_0^{-1}\langle\mathbb{M}\rangle^{-1}\langle\mathbb{M}\mathbb{P}_0[\boldsymbol{G}]\rangle \\ &= -\mathbb{M}\mathbb{P}_0[\boldsymbol{G}] + \mathbb{M}\langle\mathbb{M}\rangle^{-1}\langle\mathbb{M}\mathbb{P}_0[\boldsymbol{G}]\rangle \\ &= -\mathbb{R}[\boldsymbol{G}] + \mathbb{A}\langle\mathbb{R}[\boldsymbol{G}]\rangle. \end{aligned} \tag{A.54}$$

A.3 Orientation Tensors for the SMC Composite Samples

The orientation tensors of second- and fourth-order used as input for the mean-field homogenization are computed based on µCT scans (Pinter et al., 2018). The tensors computed for each sample are listed in Tables A.1 to A.3.

Plaque	Orientation	$\boldsymbol{N}$	$\mathbb{N}$
1D/1	0°	$\begin{pmatrix} 0.3449 & -0.0084 & 0.0023 \\ -0.0084 & 0.6325 & 0.0056 \\ 0.0023 & 0.0056 & 0.0226 \end{pmatrix}$	$\begin{pmatrix} 0.1981 & 0.1401 & 0.0067 & 0.0010 & 0.0012 & -0.0003 \\ 0.1401 & 0.4837 & 0.0087 & 0.0046 & 0.0005 & -0.0081 \\ 0.0067 & 0.0087 & 0.0072 & 0.0000 & 0.0006 & 0.0000 \\ 0.0010 & 0.0046 & 0.0000 & 0.0087 & 0.0000 & 0.0005 \\ 0.0012 & 0.0005 & 0.0006 & 0.0000 & 0.0067 & 0.0010 \\ -0.0003 & -0.0081 & 0.0000 & 0.0005 & 0.0010 & 0.1401 \end{pmatrix}$
	45°	$\begin{pmatrix} 0.4390 & -0.1164 & 0.0031 \\ -0.1164 & 0.5317 & 0.0047 \\ 0.0031 & 0.0047 & 0.0294 \end{pmatrix}$	$\begin{pmatrix} 0.2689 & 0.1606 & 0.0095 & 0.0013 & 0.0015 & -0.0523 \\ 0.1606 & 0.3623 & 0.0088 & 0.0032 & 0.0005 & -0.0631 \\ 0.0095 & 0.0088 & 0.0111 & 0.0003 & 0.0010 & -0.0011 \\ 0.0013 & 0.0032 & 0.0003 & 0.0088 & -0.0011 & 0.0005 \\ 0.0015 & 0.0005 & 0.0010 & -0.0011 & 0.0095 & 0.0013 \\ -0.0523 & -0.0631 & -0.0011 & 0.0005 & 0.0013 & 0.1606 \end{pmatrix}$
	90°	$\begin{pmatrix} 0.5744 & 0.0046 & 0.0049 \\ 0.0046 & 0.3993 & -0.0008 \\ 0.0049 & -0.0008 & 0.0263 \end{pmatrix}$	$\begin{pmatrix} 0.4144 & 0.1497 & 0.0103 & -0.0006 & 0.0031 & -0.0028 \\ 0.1497 & 0.2433 & 0.0064 & -0.0006 & 0.0008 & 0.0074 \\ 0.0103 & 0.0064 & 0.0096 & 0.0004 & 0.0010 & 0.0000 \\ -0.0006 & -0.0006 & 0.0004 & 0.0064 & 0.0000 & 0.0008 \\ 0.0031 & 0.0008 & 0.0010 & 0.0000 & 0.0103 & -0.0006 \\ -0.0028 & 0.0074 & 0.0000 & 0.0008 & -0.0006 & 0.1497 \end{pmatrix}$

Table A.1: Orientation tensors of second- and fourth-order for the samples from dataset 1D/1.

Plaque	Orientation	$\boldsymbol{N}$	$\mathbb{N}$
1D/2 HR	0°	$\begin{pmatrix} 0.4151 & 0.0307 & 0.0027 \\ 0.0307 & 0.5780 & -0.0002 \\ 0.0027 & -0.0002 & 0.0069 \end{pmatrix}$	$\begin{pmatrix} 0.2690 & 0.1432 & 0.0030 & 0.0001 & 0.0020 & 0.0225 \\ 0.1432 & 0.4316 & 0.0032 & -0.0003 & 0.0007 & 0.0079 \\ 0.0030 & 0.0032 & 0.0007 & -0.0000 & 0.0001 & 0.0002 \\ 0.0001 & -0.0003 & -0.0000 & 0.0032 & 0.0002 & 0.0007 \\ 0.0020 & 0.0007 & 0.0001 & 0.0002 & 0.0030 & 0.0001 \\ 0.0225 & 0.0079 & 0.0002 & 0.0007 & 0.0001 & 0.1432 \end{pmatrix}$
	45°	$\begin{pmatrix} 0.4833 & 0.1463 & -0.0206 \\ 0.1463 & 0.5085 & -0.0092 \\ -0.0206 & -0.0092 & 0.0082 \end{pmatrix}$	$\begin{pmatrix} 0.3429 & 0.1367 & 0.0037 & -0.0037 & -0.0143 & 0.0613 \\ 0.1367 & 0.3684 & 0.0035 & -0.0053 & -0.0059 & 0.0842 \\ 0.0037 & 0.0035 & 0.0010 & -0.0002 & -0.0005 & 0.0008 \\ -0.0037 & -0.0053 & -0.0002 & 0.0035 & 0.0008 & -0.0059 \\ -0.0143 & -0.0059 & -0.0005 & 0.0008 & 0.0037 & -0.0037 \\ 0.0613 & 0.0842 & 0.0008 & -0.0059 & -0.0037 & 0.1367 \end{pmatrix}$
	90°	$\begin{pmatrix} 0.6205 & 0.0150 & -0.0049 \\ 0.0150 & 0.3686 & -0.0071 \\ -0.0049 & -0.0071 & 0.0109 \end{pmatrix}$	$\begin{pmatrix} 0.4842 & 0.1317 & 0.0046 & -0.0019 & -0.0040 & 0.0107 \\ 0.1317 & 0.2326 & 0.0042 & -0.0049 & -0.0009 & 0.0045 \\ 0.0046 & 0.0042 & 0.0021 & -0.0003 & 0.0000 & -0.0002 \\ -0.0019 & -0.0049 & -0.0003 & 0.0042 & -0.0002 & -0.0009 \\ -0.0040 & -0.0009 & 0.0000 & -0.0002 & 0.0046 & -0.0019 \\ 0.0107 & 0.0045 & -0.0002 & -0.0009 & -0.0019 & 0.1317 \end{pmatrix}$

Table A.2: Orientation tensors of second- and fourth-order for the samples from dataset 1D/2 HR.

Plaque	Orientation	$\boldsymbol{N}$	$\mathbb{N}$
2D/1	0°	$\begin{pmatrix} 0.4997 & -0.0259 & 0.0090 \\ -0.0259 & 0.4818 & -0.0009 \\ 0.0090 & -0.0009 & 0.0185 \end{pmatrix}$	$\begin{pmatrix} 0.4159 & 0.1659 & 0.0090 & -0.0004 & 0.0072 & -0.0193 \\ 0.1659 & 0.3964 & 0.0074 & -0.0007 & 0.0027 & -0.0112 \\ 0.0090 & 0.0074 & 0.0055 & 0.0001 & 0.0008 & -0.0001 \\ -0.0004 & -0.0007 & 0.0001 & 0.0074 & -0.0001 & 0.0027 \\ 0.0072 & 0.0027 & 0.0008 & -0.0001 & 0.0090 & -0.0004 \\ -0.0193 & -0.0112 & -0.0001 & 0.0027 & -0.0004 & 0.1659 \end{pmatrix}$
	45°	$\begin{pmatrix} 0.5046 & 0.0051 & 0.0029 \\ 0.0051 & 0.4775 & 0.0022 \\ 0.0029 & 0.0022 & 0.0179 \end{pmatrix}$	$\begin{pmatrix} 0.4232 & 0.1634 & 0.0085 & 0.0008 & 0.0024 & 0.0019 \\ 0.1634 & 0.3924 & 0.0073 & 0.0015 & 0.0008 & 0.0040 \\ 0.0085 & 0.0073 & 0.0052 & 0.0004 & 0.0003 & 0.0001 \\ 0.0008 & 0.0015 & 0.0004 & 0.0073 & 0.0001 & 0.0008 \\ 0.0024 & 0.0008 & 0.0003 & 0.0001 & 0.0085 & 0.0008 \\ 0.0019 & 0.0040 & 0.0001 & 0.0008 & 0.0008 & 0.1634 \end{pmatrix}$
	90°	$\begin{pmatrix} 0.4598 & 0.0541 & -0.0005 \\ 0.0541 & 0.5104 & -0.0197 \\ -0.0005 & -0.0197 & 0.0298 \end{pmatrix}$	$\begin{pmatrix} 0.3027 & 0.1469 & 0.0102 & -0.0044 & 0.0006 & 0.0245 \\ 0.1469 & 0.3550 & 0.0085 & -0.0135 & -0.0003 & 0.0285 \\ 0.0102 & 0.0085 & 0.0111 & -0.0018 & -0.0009 & 0.0010 \\ -0.0044 & -0.0135 & -0.0018 & 0.0085 & 0.0010 & -0.0003 \\ 0.0006 & -0.0003 & -0.0009 & 0.0010 & 0.0102 & -0.0044 \\ 0.0245 & 0.0285 & 0.0010 & -0.0003 & -0.0044 & 0.1469 \end{pmatrix}$

Table A.3: Orientation tensors of second- and fourth-order for the samples from dataset 2D/1.

A.4 Settings for the Full-Field Simulations

The full-field simulations are performed with the commercial software tool GeoDict® (standard edition, version 2017) by the Math2Market GmbH. The effective elastic and thermal properties of the composite are computed using the *Elasto-Dict* module and the *FeelMath* solver. The settings and solver options for the full-field computations are listed in Table A.4.

Constituent materials	
Matrix	$E_{\mathrm{M}} = 3.4\,\mathrm{GPa}$, $\nu_{\mathrm{M}} = 0.385$, $\alpha_\theta = 7 \times 10^{-5}\,\mathrm{K}^{-1}$
Fiber	$E_{\mathrm{F}} = 73\,\mathrm{GPa}$, $\nu_{\mathrm{M}} = 0.22$, $\alpha_\theta = 0.5 \times 10^{-5}\,\mathrm{K}^{-1}$
UD-fibers	$E_1 = 31.29\,\mathrm{GPa}$, $E_2 = 8.23\,\mathrm{GPa}$, $\nu_1 = 0.45$, $\nu_2 = 0.31$, $G = 2.92\,\mathrm{GPa}$, $\alpha_\theta = 0.5 \times 10^{-5}\,\mathrm{K}^{-1}$
Load case	
Load type	Strain
Strain increment	0.05%
Boundary conditions	Periodic
Solver	
Tolerance	0.0001
Max. iterations	100 000
Method	Neumann series (memory efficient)
Parallelization	Local, 4 threads

Table A.4: Settings and solver options for the full-field simulations.

Appendix B

Production of Pure Resin Samples

For the mean-field and full-field simulations presented in Chapter 6, parameters for the pure resin material are desired. The pure resin plaques have been produced by compression molding at the Institute of Production Science (wbk). The uncured material is prior portioned to cylinder-shaped pieces at $10\,\mathrm{cm} \times 2.5\,\mathrm{cm}$, see Figure B.1a. The mold has been heated to a fixed temperature. Subsequently, each piece is positioned into the mold, cf. Figure B.1b and pressed for a specific time until a cured plaque is yielded. Since the production parameters were not clear prior to the compression molding, several trials have been performed, listed in Table B.1. As it is shown in Figure B.1c, some process parameters for the mold lead to plaques that are irregular cured and exhibit areas that are darker or brighter compared to other areas. Furthermore, voids and holes are detected within the plaque. In contrast, setting a temperature of $145\,^{\circ}\mathrm{C}$, a pressure of $2000\,\mathrm{kN}$ and a compression time of $90\,\mathrm{sec}$ as process parameters for the mold, yields plaques with greater areas of homogeneously cured material and without voids, cf. Figure B.1d. From these plaques, DMA samples were extracted and used to determined the storage modulus and the CTE value for the matrix material. Table B.1 lists the process parameters that yield successful and unsuccessful compression molding results.

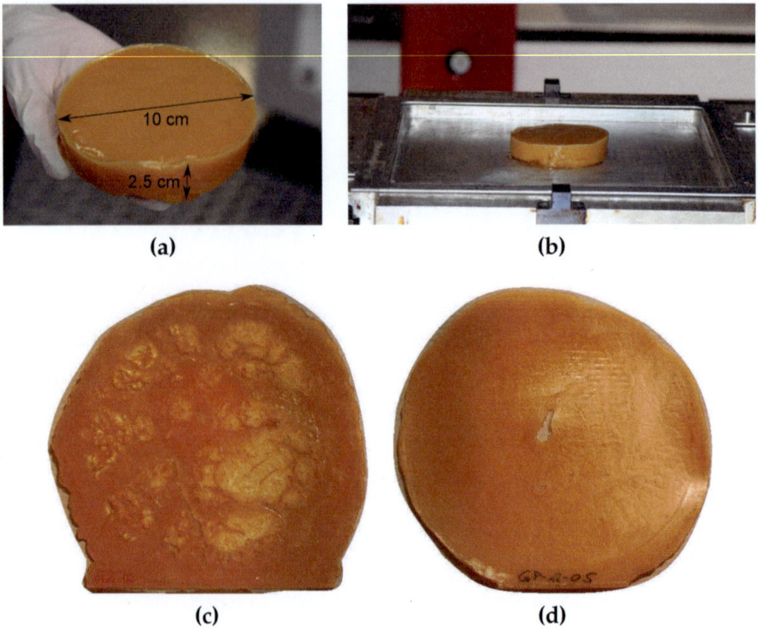

Figure B.1: a) Uncured pure UPPH piece with geometrical dimensions. b) Mold with pure resin piece. c) Cured pure resin plaque with unregular cured areas. d) Cured pure resin plaque with areas of homogeneous cured material. The pure resin plaques were manufactured at the Institute of Production Science (wbk), Production Systems.

Mold temp.	Compr. force	Compr. time	Resultant plaques
120°C - 130°C	980 kN - 1500 kN	5 min - 15 min	Unregular, heterogeneous curing, voids in the plaques
145°C	2000 kN	90 sec	Regular, homogeneous curing, nearly no voids

Table B.1: Parameters for the production by compression molding.

Appendix C

Application to Sheet Molding Compound Composite

C.1 Mean-Field Simulation Results for 1D/2 Plaque

In Section 6.2.1, mean-field results for samples from the 1D/2 plaque for $3\,\mu m$ resolutions are presented. The same samples have been scanned beforehand at a resolution of $5.5\,\mu m$. In addition to the results from Section 6.2.1, the mean-field simulation results for the coarser resultion are presented here.

The relative deviation of the orientation tensor for each sample from the planar isotropic orientation tensor is depicted in Figure C.1a. The $0°$- and $90°$-oriented samples exhibit a relative error from the planar isotropic state of approximately 20%. For the $45°$ sample, a pronounced deviation of approximately 35% is reached. In Figure C.1b, the eigenvalue tupels are shown. In particular, the $45°$ sample exhibits pronounced anisotropic properties with tendencies toward the UD fiber orientation. In contrast, the $0°$- and $90°$ samples are close to each other and exhibit a less pronounced anisotropic orientation of the microstructure compared to the $90°$ sample.

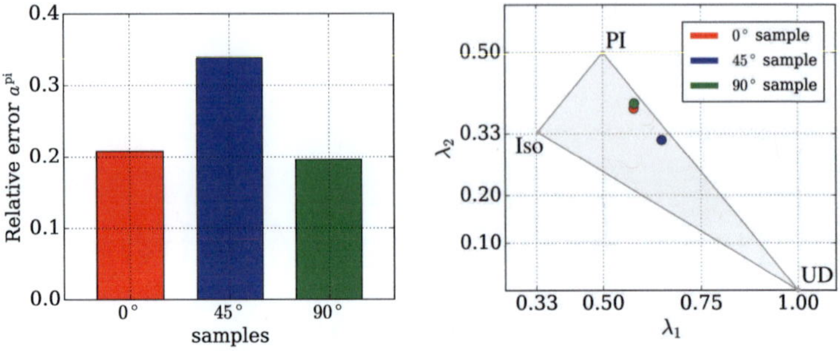

(a) **(b)**

Figure C.1: a) Relative error $a^{\text{pi}}_{\gamma} = \|\boldsymbol{N}_{\gamma} - \boldsymbol{N}^{\text{pi}}\| / \|\boldsymbol{N}^{\text{pi}}\|$ between the orientation tensors $\boldsymbol{N}_{\gamma}$ of $\alpha = \{0°, 45°, 90°\}$ samples from 1D/2 plaque and the planar isotropic reference orientation tensor $\boldsymbol{N}^{\text{pi}}$. b) Eigenvalues λ_1 and λ_2 of the orientation tenors, with $\lambda_1 \geq \lambda_2$. The abbreviations PI, Iso and UD denote planar isotropic, isotropic and unidirectional. The 45° sample exhibits a pronounced anisotropic microstructure, it is less pronounced for the 0° and 90° sample.

In Figure C.2, the mean-field results from the HS$^{\pm}$ method are shown. In the upper row, the results from HS^{+} are presented, whereas the results from the HS^{-} method are depicted in the lower row. For the HS^{+}, a rather isotropic distribution of the effective Young's modulus $\bar{E}(\boldsymbol{d})$ is computed. Slight deviations in terms of the shape of $\bar{E}(\boldsymbol{d})$ are shown for the different orientation of the samples. Furthermore, a vanishing influence of parameter k on the HS^{+} results is observed. In general, the simulation results are consistently stiffer compared to the experimental data. In contrast, the effective modulus for the HS^{-} method exhibits a pronounced anisotropic behavior. The results depend on the orientation of the sample. Furthermore, for increasing values of parameter k, the results from HS^{-} are closer to the corresponding experimental data.

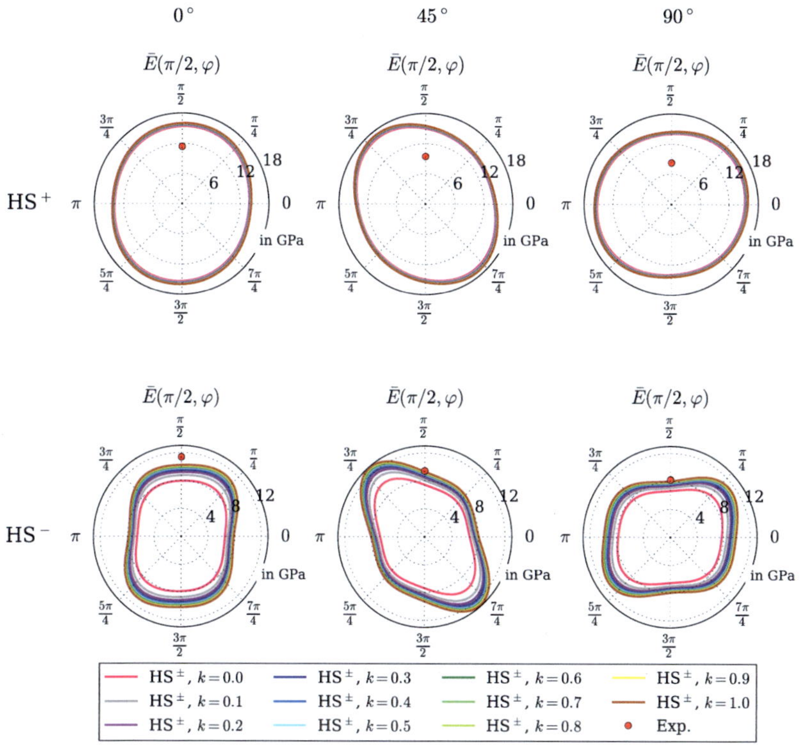

Figure C.2: Experimental data are compared to results of the HS mean-field simulation for 0°, 45° and 90° samples from 1D/2 plaque. HS^+ and HS^- results are depicted in the upper and lower row, respectively. Parameter k of the variable reference stiffness is varied. Experimental results are depicted by red dots. The effective Young's modulus is illustrated in the x-y-plane, corresponding to $\bar{E}(\pi/2, \varphi)$. Simulation results are based on a µCT scan with a resolution of 5.5 µm. Negligible sensitivity of the HS^+ method on the variation of k is observed. Significant overestimation of the experimental results by the HS^+ method is given. Results from HS^- are close to experimental data.

Figure C.3 shows the relative error between experimental data and mean-field results from the HS$^-$ method dependent on parameter k With increasing values of k, the relative error decreases monotonically. The highest deviations from the experimental data are computed for the $0°$ sample. The smallest relative error for all three samples is obtained for $k = 1.0$ in the range of 10% ($0°$) and 2% ($45°$).

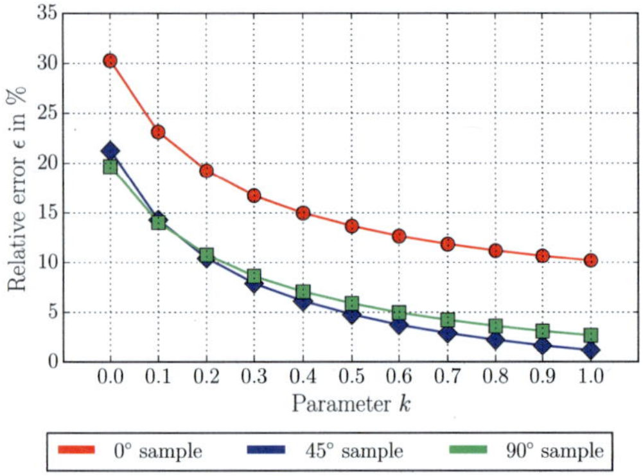

Figure C.3: Relative error $\epsilon = |E_{\text{sim}}/E_{\text{exp}} - 1.0| \cdot 100\,\%$ between experimental data and simulations for HS$^-$ is depicted for different values of k. Experimental data are obtained for $0°$-, $45°$- and $90°$ samples from the 1D/2 plaque. The relative error decreases monotonically for increasing values of k. The relative error varies between 10% ($0°$) and 2% ($45°$) for $k = 1.0$.

C.2 Mean-Field Simulation Results for $c_{\mathrm{F}} = 0.22$

For samples from the 1D/1 plaque, the mean-field simulations are performed with the input values listed in Table 6.2, except the fiber volume content is set to the nominal value of 22 vol.%.

The relative error between the average fiber orientation of each sample and the planar isotropic state is shown in Figure 6.4.

The effective Young's modulus $\bar{E}(\boldsymbol{d})$ from the HS$^{\pm}$ methods is depicted in Figure C.4. The HS^{+} results yield stiffer and rather isotropic results for $\bar{E}(\boldsymbol{d})$. From the HS^{-} method, the results show anisotropic effective behavior and are dependent on the choice of k. In comparison to experimental data, small deviations are computed.

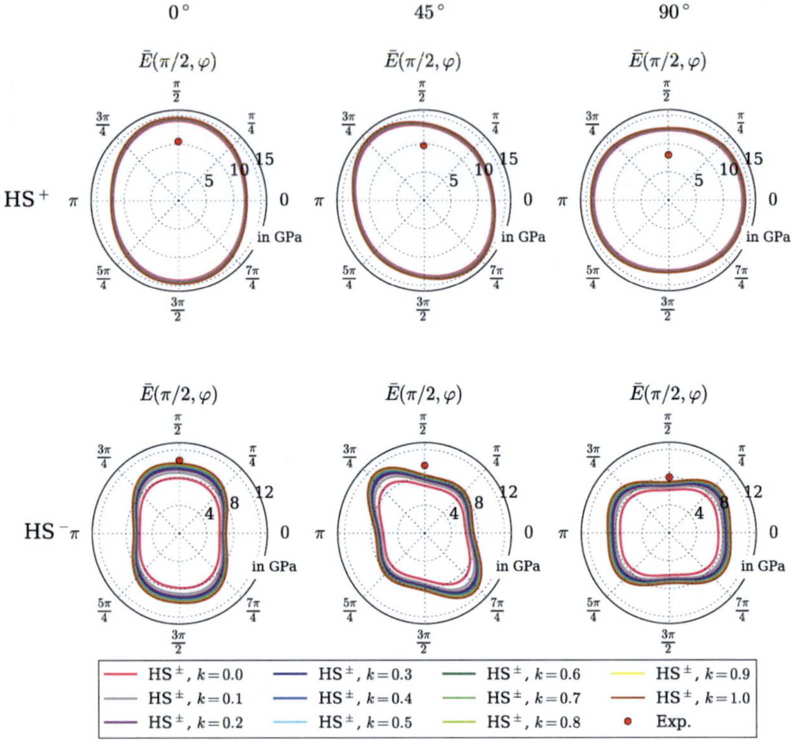

Figure C.4: Experimental data are compared to results of the HS mean-field simulation for $0°$, $45°$ and $90°$ samples from 1D/1 plaque for $c_{\mathrm{F}} = 0.22$. HS^{+} and HS^{-} results are depicted in the upper and lower row, respectively. Parameter k of the variable reference stiffness is varied. Experimental results are depicted by red dots. The effective Young's modulus is illustrated in the x-y-plane, corresponding to $\bar{E}(\pi/2, \varphi)$. Simulation results are based on a μCT scan with a resolution of 5.5 μm. Negligible sensitivity of the HS^{+} method on the variation of k is observed. Significant overestimation of the experimental results by the HS^{+} method is given. Results from HS^{-} are close to experimental data.

The relative error between simulation results for the HS^- method dependent on parameter k, and the experimental data is shown in Figure C.5. With increasing k, the relative error ϵ decreases. The smallest error is determined for $k = 1.0$ for all samples. The relative error varies between 5% (0°) and 15% (45°).

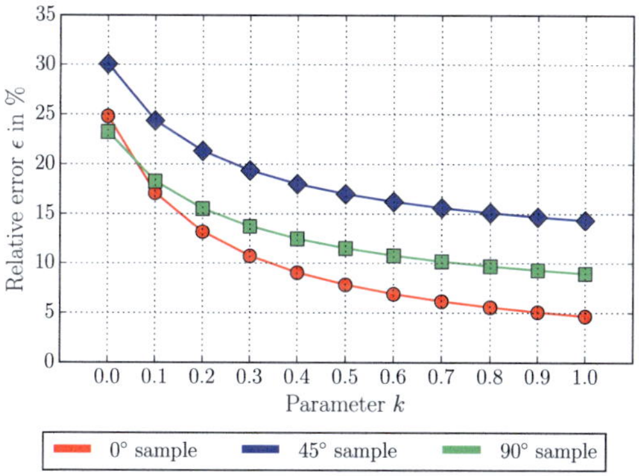

Figure C.5: Relative error $\epsilon = |E_{\text{sim}}/E_{\text{exp}} - 1.0| \cdot 100\,\%$ between experimental data and simulations for HS^- is depicted for different values of k and $c_F = 0.22$. Experimental data are obtained for 0°, 45° and 90° samples from the 1D/1 plaque. The relative error decreases monotonically for increasing values of k. The relative error varies between 15% (45°) and 5% (0°) for $k = 1.0$.

Bibliography

Abdin, Y., Lomov, S. V., Jain, A., Van Lenthe, G., Verpoest, I., 2014. Geometrical characterization and micro-structural modeling of short steel fiber composites. Composites Part A: Applied Science and Manufacturing 67, 171–180.

Advani, S. G., Tucker III, C. L., 1987. The use of tensors to describe and predict fiber orientation in short fiber composites. Journal of Rheology 31 (8), 751–784.

Akay, M., 1993. Aspects of dynamic mechanical analysis in polymeric composites. Composites Science and Technology 47 (4), 419–423.

Albers, A., Spadinger, M., Serf, M., Reichert, S., Heldmaier, S., Schulz, M., Bursac, N., 2017. Coupling of computer-aided methods: Supporting product developer during embodiment synthesis. In: World Congress of Structural and Multidisciplinary Optimisation. Springer, pp. 536–548.

Anagnostou, D., Chatzigeorgiou, G., Chemisky, Y., Meraghni, F., 2018. Hierarchical micromechanical modeling of the viscoelastic behavior coupled to damage in SMC and SMC-hybrid composites. Composites Part B: Engineering 151, 8–24.

Andrä, H., Combaret, N., Dvorkin, J., Glatt, E., Han, J., Kabel, M., Keehm, Y., Krzikalla, F., Lee, M., Madonna, C., et al., 2013. Digital rock physics benchmarks - Part II: Computing effective properties. Computers & Geosciences 50, 33–43.

Bay, R. S., Tucker III, C. L., 1992. Fiber orientation in simple injection moldings. Part I: Theory and numerical methods. Polymer Composites 13 (4), 317–331.

Beegle, B. L., Modell, M., Reid, R. C., 1974. Legendre transforms and their application in thermodynamics. American Institute of Chemical Engineers Journal 20 (6), 1194–1200.

Benveniste, Y., 1987. A new approach to the application of Mori-Tanaka's theory in composite materials. Mechanics of Materials 6 (2), 147–157.

Benveniste, Y., 2008. Revisiting the generalized self-consistent scheme in composites: Clarification of some aspects and a new formulation. Journal of the Mechanics and Physics of Solids 56 (10), 2984–3002.

Benveniste, Y., Dvorak, G., Chen, T., 1989. Stress fields in composites with coated inclusions. Mechanics of Materials 7 (4), 305–317.

Bernasconi, A., Cosmi, F., Dreossi, D., 2008. Local anisotropy analysis of injection moulded fibre reinforced polymer composites. Composites Science and Technology 68 (12), 2574–2581.

Bernasconi, A., Cosmi, F., Hine, P., 2012. Analysis of fibre orientation distribution in short fibre reinforced polymers: A comparison between optical and tomographic methods. Composites Science and Technology 72 (16), 2002–2008.

Bertóti, R., Böhlke, T., 2017. Flow-induced anisotropic viscosity in short FRPs. Mechanics of Advanced Materials and Modern Processes 3, 1–12.

Bertram, A., 2005. Elasticity and Plasticity of Large Deformations. Springer, Heidelberg.

Bertram, A., Krawietz, A., 2012. On the introduction of thermoplasticity. Acta Mechanica 223 (10), 2257–2268.

Bishop, J., Hill, R., 1951a. A theoretical derivation of the plastic properties of a polycrystalline face-centred metal. The London, Edinburgh, and Dublin Philosophical Magazine and Journal of Science 42 (334), 1298–1307.

Bishop, J., Hill, R., 1951b. A theory of the plastic distortion of a polycrystalline aggregate under combined stresses. The London, Edinburgh, and Dublin Philosophical Magazine and Journal of Science 42 (327), 414–427.

Böhlke, T., Brüggemann, C., 2001. Graphical representation of the generalized Hooke's law. Technische Mechanik 21 (2), 145–158.

Böhlke, T., Lobos, M., 2014. Representation of Hashin–Shtrikman bounds of cubic crystal aggregates in terms of texture coefficients with application in materials design. Acta Materialia 67, 324–334.

Böhm, H. J., 2004. Mechanics of Microstructured Materials. Vol. 464. Springer, Vienna.

Brassart, L., Inglis, H., Delannay, L., Doghri, I., Geubelle, P., 2009. An extended Mori–Tanaka homogenization scheme for finite strain modeling of debonding in particle-reinforced elastomers. Computational Materials Science 45 (3), 611–616.

Brinson, H. F., Brinson, L. C., 2015. Polymer Engineering Science and Viscoelasticity: An Introduction, 2nd Edition. Springer, New York, Heidelberg, Dordrecht, London.

Brylka, B., 2017. Charakterisierung und Modellierung der Steifigkeit von langfaserverstärktem Polypropylen. Doctoral Disseration, KIT Scientific Publishing, Schriftenreihe Kontinuumsmechanik im Maschinenbau Nr. 10.

Brylka, B., Schemmann, M., Wood, J., Böhlke, T., 2018. DMA based characterization of stiffness reduction in long fiber reinforced polypropylene. Polymer Testing 66, 296–302.

Budiansky, B., 1965. On the elastic moduli of some heterogeneous materials. Journal of the Mechanics and Physics of Solids 13 (4), 223–227.

Busfield, J., Muhr, A., 2003. Constitutive models for rubber III. In: Proceedings of the Third European Conference on Constitutive Models for Rubber. CRC Press, pp. 333 –341.

Bücheler, D., 2018. Locally continuous-fiber reinforced sheet molding compound. Doctoral Dissertation, Fraunhofer Verlag, Wissenschaftliche Schriftenreihe des Fraunhofer ICT Band 79.

Camacho, C. W., Tucker, C. L., Yalvaç, S., McGee, R. L., 1990. Stiffness and thermal expansion predictions for hybrid short fiber composites. Polymer Composites 11 (4), 229–239.

Carus, M., 2011. Bio-composites: Technologies, applications and markets. In: 4th International Conference on Sustainable Materials, Polymers and Composites. pp. 6–7.

Castañeda, P. P., 1991. The effective mechanical properties of nonlinear isotropic composites. Journal of the Mechanics and Physics of Solids 39 (1), 45–71.

Castañeda, P. P., Suquet, P., 1998. Nonlinear composites. In: Advances in Applied Mechanics. Vol. 34. Elsevier, pp. 171–302.

Chen, T., Dvorak, G., Benveniste, Y., 1990. Stress fields in composites reinforced by coated cylindrically orthotropic fibers. Mechanics of Materials 9 (1), 17–32.

Christensen, R., Lo, K., 1979. Solutions for effective shear properties in three phase sphere and cylinder models. Journal of the Mechanics and Physics of Solids 27 (4), 315–330.

Clarke, A., Davidson, N., Archenhold, G., 1993. Measurements of fibre direction in reinforced polymer composites. Journal of Microscopy 171 (1), 69–79.

Coleman, B. D., Gurtin, M. E., 1967. Thermodynamics with internal state variables. The Journal of Chemical Physics 47 (2), 597–613.

Coleman, B. D., Noll, W., 1963. The thermodynamics of elastic materials with heat conduction and viscosity. Archive for Rational Mechanics and Analysis 13 (1), 167–178.

Cverna, F., 2002. ASM Ready Reference: Thermal Properties of Metals. ASM International, Materials Park, Ohio.

Daniels, F., ter Haar Romeny, van Assen, Strijkers, G., Engels, W., Rubbens, M., 2006. Quantification of collagen orientation in 3D engineered tissue. In: 3rd International Conference on Biomedical Engineering. Vol. 15. Springer, pp. 282–286.

Denault, J., Vu-Khanh, T., Foster, B., 1989. Tensile properties of injection molded long fiber thermoplastic composites. Polymer Composites 10 (5), 313–321.

Dlubek, G., Gupta, A. S., Pionteck, J., Häßler, R., Krause-Rehberg, R., Kaspar, H., Lochhaas, K., 2005. Glass transition and free volume in the mobile (MAF) and rigid (RAF) amorphous fractions of semicrystalline PTFE: a positron lifetime and PVT study. Polymer 46 (16), 6075–6089.

Doghri, I., Friebel, C., 2005. Effective elasto-plastic properties of inclusion-reinforced composites. Study of shape, orientation and cyclic response. Mechanics of Materials 37 (1), 45–68.

Doghri, I., Tinel, L., 2005. Micromechanical modeling and computation of elasto-plastic materials reinforced with distributed-orientation fibers. International Journal of Plasticity 21 (10), 1919–1940.

Drugan, W., Willis, J., 1996. A micromechanics-based nonlocal constitutive equation and estimates of representative volume element size for elastic composites. Journal of the Mechanics and Physics of Solids 44 (4), 497–524.

Du, D.-X., Zheng, Q.-S., 2002. A further exploration of the interaction direct derivative (IDD) estimate for the effective properties of multiphase composites taking into account inclusion distribution. Acta Mechanica 157 (1-4), 61–80.

Düster, A., Sehlhorst, H.-G., Rank, E., 2012. Numerical homogenization of heterogeneous and cellular materials utilizing the finite cell method. Computational Mechanics 50 (4), 413–431.

Dvorak, G., 2012. Micromechanics of Composite Materials. Solid Mechanics and Its Applications. Springer, Dordrecht, Heidelberg, New York, London.

Ehrenstein, G., 1999. Polymer-Werkstoffe, 2nd Edition. Studientexte Kunststofftechnik. Hanser-Verlag, München.

Einstein, A., 1906. Eine neue Bestimmung der Moleküldimensionen. Annalen der Physik 324 (2), 289–306.

Eisenlohr, P., Diehl, M., Lebensohn, R. A., Roters, F., 2013. A spectral method solution to crystal elasto-viscoplasticity at finite strains. International Journal of Plasticity 46, 37–53.

Eshelby, J. D., 1957. The determination of the elastic field of an ellipsoidal inclusion, and related problems. Proceedings of the Royal Society A 241 (1226), 376–396.

European Union, July 2018. Reducing CO2 emissions from passenger cars.
URL `https://ec.europa.eu/clima/policies/transport/vehicles/cars_en`

Fakirov, S., Fakirova, C., 1985. Direct determination of the orientation of short glass fibers in an injection-molded poly (ethylene terephthalate) system. Polymer Composites 6 (1), 41–46.

Feldkamp, L., Davis, L., Kress, J., 1984. Practical cone-beam algorithm. Journal of the Optical Society of America A, Optics and Image Science 1 (6), 612–619.

Fengler, B., Kärger, L., Henning, F., Hrymak, A., 2018. Multi-objective patch optimization with integrated kinematic draping simulation for continuous–discontinuous fiber-reinforced composite structures. Journal of Composites Science 2 (2), 1–19.

Ferry, J. D., 1980. Viscoelastic Properties of Polymers, 3rd Edition. Wiley, New York, Chichester, Brisbane, Toronto, Singapore.

Fliegener, S., 2015. Micromechanical finite element modeling of long fiber reinforced thermoplastics. Doctoral Dissertation, Karlsruher Institut für Technologie (KIT).

Fung, K., Li, R. K., 2006. Mechanical properties of short glass fibre reinforced and functionalized rubber-toughened pet blends. Polymer Testing 25 (7), 923–931.

Gabbott, P., 2008. Principles and Applications of Thermal Analysis, 1st Edition. Blackwell Publishing, Oxford.

Garesci, F., Fliegener, S., 2013. Young's modulus prediction of long fiber reinforced thermoplastics. Composites Science and Technology 85, 142–147.

Gillham, J., Enns, J., 1994. On the cure and properties of thermosetting polymers using torsional braid analysis. Trends in Polymer Science 2 (12), 406–418.

Goldberg, N., Ospald, F., Schneider, M., 2017. A fiber orientation-adapted integration scheme for computing the hyperelastic Tucker average for short fiber reinforced composites. Computational Mechanics 60 (4), 595–611.

Granta Design, 2016. CES Edupack materials selector.
URL `https://www.grantadesign.com/de/products/ces/`

Hall, C., 1981. Polymer Materials: An Introduction for Technologists and Scientists, 1st Edition. The Macmillan Press LTD, London.

Hashin, Z., Shtrikman, S., 1962. A variational approach to the theory of the elastic behaviour of polycrystals. Journal of the Mechanics and Physics of Solids 10 (4), 343–352.

Haupt, P., 2002. Continuum Mechanics and Theory of Materials, 2nd Edition. Advanced Texts in Physics. Springer, Berlin.

Henning, F., Moeller, E., 2011. Handbuch Leichtbau: Methoden, Werkstoffe, Fertigung, 1st Edition. Hanser-Verlag, München.

Hershey, A. V., 1954. The elasticity of an isotropic aggregate of anisotropic cubic crystals. Journal of Applied Mechanics-Transactions of the ASME 21 (3), 236–240.

Hill, R., 1952. The elastic behaviour of a crystalline aggregate. Proceedings of the Physical Society. Section A 65 (5), 349–354.

Hill, R., 1963. Elastic properties of reinforced solids: some theoretical principles. Journal of the Mechanics and Physics of Solids 11 (5), 357–372.

Hill, R., 1965. A self-consistent mechanics of composite materials. Journal of the Mechanics and Physics of Solids 13 (4), 213–222.

Hine, P., Davidson, N., Duckett, R., Ward, I., 1995. Measuring the fibre orientation and modelling the elastic properties of injection-moulded long-glass-fibre-reinforced nylon. Composites Science and Technology 53 (2), 125–131.

Hine, P., Parveen, B., Brands, D., Caton-Rose, F., 2014. Validation of the modified rule of mixtures using a combination of fibre orientation and fibre length measurements. Composites Part A: Applied science and manufacturing 64, 70–78.

Hine, P. J., Lusti, H. R., Gusev, A. A., 2002. Numerical simulation of the effects of volume fraction, aspect ratio and fibre length distribution on the elastic and thermoelastic properties of short fibre composites. Composites Science and Technology 62 (10-11), 1445–1453.

Hohberg, M., Kärger, L., Henning, F., Hrymak, A., 2017. Rheological measurements and rheological shell model considering the compressible behavior of long fiber reinforced sheet molding compound (SMC). Composites Part A: Applied Science and Manufacturing 95, 110–117.

Idiart, M., Castañeda, P. P., 2005. Second-order estimates for nonlinear isotropic composites with spherical pores and rigid particles. Comptes Rendus Mécanique 333 (2), 147–154.

Jendli, Z., Meraghni, F., Fitoussi, J., Baptiste, D., 2009. Multi-scales modelling of dynamic behaviour for discontinuous fibre SMC composites. Composites Science and Technology 69 (1), 97–103.

Jones, E., Oliphant, T., Peterson, P., et al., 2001. SciPy: Open source scientific tools for Python.
URL `http://www.scipy.org/`

Kabel, M., Böhlke, T., Schneider, M., 2014. Efficient fixed point and Newton–Krylov solvers for FFT-based homogenization of elasticity at large deformations. Computational Mechanics 54 (6), 1497–1514.

Kabel, M., Fink, A., Schneider, M., 2017. The composite voxel technique for inelastic problems. Computer Methods in Applied Mechanics and Engineering 322, 396–418.

Kabel, M., Merkert, D., Schneider, M., 2015. Use of composite voxels in FFT-based homogenization. Computer Methods in Applied Mechanics and Engineering 294, 168–188.

Kak, A., Slaney, M., Wang, G., 2002. Principles of computerized tomographic imaging. Medical Physics 29 (1), 49–112.

Kanatani, K.-I., 1984. Distribution of directional data and fabric tensors. International Journal of Engineering Science 22 (2), 149–164.

Kehrer, L., Wicht, D., Wood, J. T., Böhlke, T., 2018. Dynamic mechanical analysis of pure and fiber-reinforced thermoset-and thermoplastic-based polymers and free volume-based viscoelastic modeling. GAMM-Mitteilungen 41 (1), 1–16.

Komori, T., Marishima, K., 1978. Estimation of fiber orientation and length in fiber assemblies. Textile Research Journal 48 (6), 309–314.

Kostka, P., Holeczek, K., Höhne, R., Filippatos, A., Modler, N., 2016. Extension and application of dynamic mechanical analysis for the estimation of spatial distribution of material properties. Polymer Testing 52, 184–191.

Krause, M., Hausherr, J.-M., Burgeth, B., Herrmann, C., Krenkel, W., 2010. Determination of the fibre orientation in composites using the structure tensor and local X-ray transform. Journal of Materials Science 45 (4), 888–896.

Kröner, E., 1958. Berechnung der elastischen Konstanten des Vielkristalls aus den Konstanten des Einkristalls. Zeitschrift für Physik 151 (4), 504–518.

Kröner, E., 1977. Bounds for effective elastic moduli of disordered materials. Journal of the Mechanics and Physics of Solids 25 (2), 137–155.

Lahellec, N., Michel, J.-C., Moulinec, H., Suquet, P., 2003. Analysis of inhomogeneous materials at large strains using fast Fourier transforms. In: IUTAM symposium on computational mechanics of solid materials at large strains. Springer, pp. 247–258.

Laws, N., 1973. On the thermostatics of composite materials. Journal of the Mechanics and Physics of Solids 21 (1), 9–17.

Laws, N., McLaughlin, R., 1978. Self-consistent estimates for the viscoelastic creep compliances of composite materials. Proceedings of the Royal Society of London Series A 359 (1697), 251–273.

Lee, K. S., Lee, S. W., Chung, K., Kang, T. J., Youn, J. R., 2003. Measurement and numerical simulation of three-dimensional fiber orientation states in injection-molded short-fiber-reinforced plastics. Journal of Applied Polymer Science 88 (2), 500–509.

Levin, V., 1976. Determination of composite material elastic and thermoelastic constants. Mechanics of Solids 11 (6), 119–126.

Lielens, G., Pirotte, P., Couniot, A., Dupret, F., Keunings, R., 1998. Prediction of thermo-mechanical properties for compression moulded composites. Composites Part A: Applied Science and Manufacturing 29 (1-2), 63–70.

Liu, Y., Castañeda, P. P., 2004. Second-order theory for the effective behavior and field fluctuations in viscoplastic polycrystals. Journal of the Mechanics and Physics of Solids 52 (2), 467–495.

Lobo, H., Bonilla, J. V., 2003. Handbook of Plastics Analysis. Vol. 68. Marcel Dekker Inc., New York.

Lobos Fernández, M., 2018. Homogenization and materials design of mechanical properties of textured materials based on zeroth-, first-and second-order bounds of linear behavior. Doctoral Dissertation, KIT Scientific Publishing, Schriftenreihe Kontinuumsmechanik im Maschinenbau Nr. 12.

Maire, E., Withers, P. J., 2014. Quantitative X-ray tomography. International Materials Reviews 59 (1), 1–43.

Mandel, J., 1965. Généralisation de la théorie de plasticité de WT Koiter. International Journal of Solids and Structures 1 (3), 273–295.

Math2Market, 2017. Geodict.
URL https://www.math2market.com/?language=de/

Matzenmiller, A., Gerlach, S., 2004. Micromechanical modeling of viscoelastic composites with compliant fiber–matrix bonding. Computational Materials Science 29 (3), 283–300.

Maxwell, J., 1891. Electricity and Magnetism, 3rd Edition. Oxford University Press, Oxford.

Menard, K. P., 2008. Dynamic Mechanical Analysis: A Practical Introduction, 2nd Edition. CRC press, Boca Raton.

Meraghni, F., Benzeggagh, M., 1995. Micromechanical modelling of matrix degradation in randomly oriented discontinuous-fibre composites. Composites Science and Technology 55 (2), 171–186.

Michel, J., Moulinec, H., Suquet, P., 2001. A computational scheme for linear and non-linear composites with arbitrary phase contrast. International Journal for Numerical Methods in Engineering 52 (1-2), 139–160.

Michel, J.-C., Moulinec, H., Suquet, P., 1999. Effective properties of composite materials with periodic microstructure: a computational approach. Computer Methods in Applied Mechanics and Engineering 172 (1-4), 109–143.

Mijović, J., Liang, R., 1984. The effect of pressure and temperature on time-dependent changes in graphite/epoxy composites below the glass transition. Polymer Engineering & Science 24 (1), 57–66.

Milton, G., 1985. The coherent potential approximation is a realizable effective medium scheme. Communications in Mathematical Physics 99 (4), 463–500.

Milton, G., 2002. The Theory of Composites. Cambridge Monographs on Applied and Computational Mathematics. Cambridge University Press, Cambridge, UK.

Mlekusch, B., 1999. Thermoelastic properties of short-fibre-reinforced thermoplastics. Composites Science and Technology 59 (6), 911–923.

Monchiet, V., Bonnet, G., 2013. Numerical homogenization of nonlinear composites with a polarization-based FFT iterative scheme. Computational Materials Science 79, 276–283.

Mori, T., Tanaka, K., 1973. Average stress in matrix and average elastic energy of materials with misfitting inclusions. Acta Metallurgica 21 (5), 571–574.

Moulinec, H., Suquet, P., 1994. A fast numerical method for computing the linear and nonlinear mechanical properties of composites. Comptes Rendus de l' Academie des Sciences Serie IIb: Mecanique Physique Chimie Astronomie 318 (11), 1417–1423.

Moulinec, H., Suquet, P., 1998. A numerical method for computing the overall response of nonlinear composites with complex microstructure. Computer Methods in Applied Mechanics and Engineering 157 (1-2), 69–94.

Mouzakis, D. E., Zoga, H., Galiotis, C., 2008. Accelerated environmental ageing study of polyester/glass fiber reinforced composites (GFRPCs). Composites Part B: Engineering 39 (3), 467–475.

Müller, I., 1985. Thermodynamics. Interaction of Mechanics and Mathematics Series. Pitman, Boston.

Müller, V., 2016. Micromechanical modeling of short-fiber reinforced composites. Doctoral Dissertation, KIT Scientific Publishing, Schriftenreihe Kontinuumsmechanik im Maschinenbau Nr. 6.

Müller, V., Brylka, B., Dillenberger, F., Glöckner, R., Kolling, S., Böhlke, T., 2016. Homogenization of elastic properties of short-fiber reinforced composites based on measured microstructure data. Journal of Composite Materials 50 (3), 297–312.

Nemat-Nasser, S., Hori, M., 1993. Micromechanics: Overall Properties of Heterogeneous Elastic Solids. Vol. 37. North-Holland Series in Applied Mathematics and Mechanics, Amsterdam, London, New York, Tokyo.

Nguyen, B. N., Bapanapalli, S. K., Holbery, J. D., Smith, M. T., Kunc, V., Frame, B. J., Phelps, J. H., Tucker III, C. L., 2008. Fiber length and orientation in long-fiber injection-molded thermoplastics–Part I: Modeling of microstructure and elastic properties. Journal of Composite Materials 42 (10), 1003–1029.

Nix, F., MacNair, D., 1941. The thermal expansion of pure metals: copper, gold, aluminum, nickel, and iron. Physical Review 60 (8), 597–605.

Ohser, J., Schladitz, K., 2009. 3D Images of Materials Structures: Processing and Analysis. John Wiley & Sons, Weinheim.

Oliphant, T. E., 2006. A Guide to NumPy. Vol. 1. Trelgol Publishing, USA.

Ostoja-Starzewski, M., 2006. Material spatial randomness: From statistical to representative volume element. Probabilistic Engineering Mechanics 21 (2), 112–132.

Phelps, J. H., Tucker III, C. L., 2009. An anisotropic rotary diffusion model for fiber orientation in short-and long-fiber thermoplastics. Journal of Non-Newtonian Fluid Mechanics 156 (3), 165–176.

Pierard, O., Friebel, C., Doghri, I., 2004. Mean-field homogenization of multi-phase thermo-elastic composites: a general framework and its validation. Composites Science and Technology 64 (1011), 1587–1603.

Pinter, P., 2018. Microstructure characterisation of continuous-discontinuous fibre reinforced polymers based on volumetric images. Doctoral Dissertation, Karlsruhe Institute of Technology (KIT).

Pinter, P., Dietrich, S., Bertram, B., Kehrer, L., Elsner, P., Weidenmann, K., 2018. Comparison and error estimation of 3d fibre orientation analysis of computed tomography image data for fibre reinforced composites. NDT & E International 95, 26–35.

Poynting, J. H., et al., 1909. On pressure perpendicular to the shear planes in finite pure shears, and on the lengthening of loaded wires when twisted. Proceedings of the Royal Society of London Series A 82 (557), 546–559.

Reddy, N., 2015. A review on completely biodegradable composites developed using soy-based matrices. Journal of Reinforced Plastics and Composites 34 (18), 1457–1475.

Reed, K. E., 1980. Dynamic mechanical analysis of fiber reinforced composites. Polymer Composites 1 (1), 44–49.

Reuss, A., 1929. Berechnung der Fließgrenze von Mischkristallen auf Grund der Plastizitätsbedingung für Einkristalle. Journal of Applied Mathematics and Mechanics 9 (1), 49–58.

Robb, K., Wirjadi, O., Schladitz, K., 2007. Fiber orientation estimation from 3d image data: Practical algorithms, visualization, and interpretation. In: 7th International Conference on Hybrid Intelligent Systems. Institute of Electrical and Electronics Engineers (IEEE), pp. 320–325.

Schapery, R., 1967. Stress analysis of viscoelastic composite materials. Journal of Composite Materials 1 (3), 228–267.

Schapery, R. A., 1968. Thermal expansion coefficients of composite materials based on energy principles. Journal of Composite Materials 2 (3), 380–404.

Schemmann, M., 2018. Biaxial characterization and mean-field based damage modeling of sheet molding compound composites. Doctoral Dissertation, KIT Scientific Publishing, Schriftenreihe Kontinuumsmechanik im Maschinenbau Nr. 13.

Schemmann, M., Gajek, S., Böhlke, T., 2018a. Biaxial tensile tests and microstructure-based inverse parameter identification of inhomogeneous SMC composites. In: Advances in Mechanics of Materials and Structural Analysis. Springer, pp. 329–342.

Schemmann, M., Görthofer, J., Seelig, T., Hrymak, A., Böhlke, T., 2018b. Anisotropic meanfield modeling of debonding and matrix damage in SMC composites. Composites Science and Technology 161, 143–158.

Schemmann, M., Lang, J., Helfrich, A., Seelig, T., Böhlke, T., 2018c. Cruciform specimen design for biaxial tensile testing of SMC. Journal of Composites Science 2 (12), 1–21.

Schneider, D., Schwab, F., Schoof, E., Reiter, A., Herrmann, C., Selzer, M., Böhlke, T., Nestler, B., 2017a. On the stress calculation within phase-field approaches: a model for finite deformations. Computational Mechanics 60 (2), 203–217.

Schneider, M., 2017a. An FFT-based fast gradient method for elastic and inelastic unit cell homogenization problems. Computer Methods in Applied Mechanics and Engineering 315, 846–866.

Schneider, M., 2017b. The sequential addition and migration method to generate representative volume elements for the homogenization of short fiber reinforced plastics. Computational Mechanics 59 (2), 247–263.

Schneider, M., Merkert, D., Kabel, M., 2017b. FFT-based homogenization for microstructures discretized by linear hexahedral elements. International Journal for Numerical Methods in Engineering 109 (10), 1461–1489.

Schneider, M., Ospald, F., Kabel, M., 2016. Computational homogenization of elasticity on a staggered grid. International Journal for Numerical Methods in Engineering 105 (9), 693–720.

Schober, M., Kuboki, T., Ameri, E., Hohe, J., Peter, G., 2017. Effects of process parameters on the interlaminar fracture toughness of GF-PA6-tapes. Proceeding in Applied Mathematics and Mechanics 17 (1), 273–274.

Shen, H., Nutt, S., Hull, D., 2004. Direct observation and measurement of fiber architecture in short fiber-polymer composite foam through micro-CT imaging. Composites Science and Technology 64 (13-14), 2113–2120.

Sliseris, J., Andrä, H., Kabel, M., Dix, B., Plinke, B., Wirjadi, O., Frolovs, G., 2014. Numerical prediction of the stiffness and strength of medium density fiberboards. Mechanics of Materials 79, 73–84.

Spahn, J., Andrä, H., Kabel, M., Müller, R., 2014. A multiscale approach for modeling progressive damage of composite materials using fast Fourier transforms. Computer Methods in Applied Mechanics and Engineering 268, 871–883.

Talbot, D., Willis, J., 1985. Variational principles for inhomogeneous non-linear media. IMA Journal of Applied Mathematics 35 (1), 39–54.

Thomason, J., 1990. Investigation of composite interphase using dynamic mechanical analysis: artifacts and reality. Polymer Composites 11 (2), 105–113.

Tjong, S. C., Xu, S.-A., Li, R. K.-Y., Mai, Y.-W., 2002. Mechanical behavior and fracture toughness evaluation of maleic anhydride compatibilized short glass fiber/sebs/polypropylene hybrid composites. Composites Science and Technology 62 (6), 831–840.

Torquato, S., 2002. Random Heterogeneous Materials: Microstructure and Macroscopic Properties. Springer, New York.

Touloukian, Y. S., Ho, C., 1970. Thermal Expansion. Metallic Elements and Alloys. Thermophysical Properties of Matter. TPRC Data Books, New York.

Trauth, A., Bondy, M., Weidenmann, K., Altenhof, W., 2018. Mechanical properties and damage evolution of a structural sheet molding compound based on a novel two step curing resin system. Materials & Design 143, 224–237.

Trauth, A., Weidenmann, K. A., 2018. Continuous-discontinuous sheet moulding compounds–Effect of hybridisation on mechanical material properties. Composite Structures 202, 1087–1098.

Truesdell, C., Toupin, R., 1960. The Classical Field Theories. Springer, Berlin, Heidelberg, Wien.

Tucker III, C. L., Liang, E., 1999. Stiffness predictions for unidirectional short-fiber composites: review and evaluation. Composites Science and Technology 59 (5), 655–671.

Vinogradov, V., Milton, G., 2008. An accelerated FFT algorithm for thermoelastic and non-linear composites. International Journal for Numerical Methods in Engineering 76 (11), 1678–1695.

Voigt, W., 1889. Ueber die Beziehung zwischen den beiden Elasticitätsconstanten isotroper Körper. Annalen der Physik 274 (12), 573–587.

Walpole, L., 1966. On bounds for the overall elastic moduli of inhomogeneous systems–I. Journal of the Mechanics and Physics of Solids 14 (3), 151–162.

Walpole, L., 1969. On the overall elastic moduli of composite materials. Journal of the Mechanics and Physics of Solids 17 (4), 235–251.

Wang, H., Zhou, H., Mishnaevsky, L., Brøndsted, P., Wang, L., 2009. Single fibre and multifibre unit cell analysis of strength and cracking of unidirectional composites. Computational Materials Science 46 (4), 810–820.

Węglewski, W., Bochenek, K., Basista, M., Schubert, T., Jehring, U., Litniewski, J., Mackiewicz, S., 2013. Comparative assessment of Young's modulus measurements of metal–ceramic composites using mechanical and non-destructive tests and micro-CT based computational modeling. Computational Materials Science 77, 19–30.

Willis, J., 1977. Bounds and self-consistent estimates for the overall properties of anisotropic composites. Journal of the Mechanics and Physics of Solids 25 (3), 185–202.

Willis, J., 1986. Variational estimates for the overall response of an inhomogeneous nonlinear dielectric. In: Homogenization and effective moduli of materials and media. Springer, pp. 247–263.

Willis, J. R., 1981. Variational and related methods for the overall properties of composites. Advances in Applied Mechanics 21, 1–78.

Willis, J. R., 1983. The overall elastic response of composite materials. Journal of Applied Mechanics 50 (4b), 1202–1209.

Witten, E., Sauer, M., Kühnel, M., 2017. Composites Market Report 2017-Market Developments, Trends, Outlook and Challenges. Tech. rep., AVK Industrievereinigung verstärkte Kunststoffe, Carbon Composites eV.

Zaiß, M., Jank, M.-H., Netzelmann, U., Waschkies, T., Rabe, U., Herrmann, H.-G., Thompson, M., Lanza, G., 2017. Use of thermography and ultrasound for the quality control of SMC lightweight material reinforced by carbon fiber tapes. Procedia CIRP 62, 33–38.

Zhao, Y., Tandon, G., Weng, G., 1989. Elastic moduli for a class of porous materials. Acta Mechanica 76 (1-2), 105–131.

Zheng, Q.-S., Du, D.-X., 2001. An explicit and universally applicable estimate for the effective properties of multiphase composites which accounts for inclusion distribution. Journal of the Mechanics and Physics of Solids 49 (11), 2765–2788.

Zheng, Q.-s., Zou, W.-n., 2001. Orientation distribution functions for microstructures of heterogeneous materials (I) – directional distribution functions and irreducible tensors. Applied Mathematics and Mechanics 22 (8), 865–884.

Schriftenreihe Kontinuumsmechanik im Maschinenbau
Karlsruher Institut für Technologie (KIT)
(ISSN 2192-693X)

Herausgeber: Prof. Dr.-Ing. Thomas Böhlke

Band 1 Felix Fritzen
Microstructural modeling and computational homogenization of the physically linear and nonlinear constitutive behavior of micro-heterogeneous materials. 2011
ISBN 978-3-86644-699-1

Band 2 Rumena Tsotsova
Texturbasierte Modellierung anisotroper Fließpotentiale. 2012
ISBN 978-3-86644-764-6

Band 3 Johannes Wippler
Micromechanical finite element simulations of crack propagation in silicon nitride. 2012
ISBN 978-3-86644-818-6

Band 4 Katja Jöchen
Homogenization of the linear and non-linear mechanical behavior of polycrystals. 2013
ISBN 978-3-86644-971-8

Band 5 Stephan Wulfinghoff
Numerically Efficient Gradient Crystal Plasticity with a Grain Boundary Yield Criterion and Dislocation-based Work-Hardening. 2014
ISBN 978-3-7315-0245-6

Band 6 Viktor Müller
Micromechanical modeling of short-fiber reinforced composites. 2016
ISBN 978-3-7315-0454-2

Die Bände sind unter www.ksp.kit.edu als PDF frei verfügbar oder als Druckausgabe bestellbar.

Band 7 Florian Rieger
Work-hardening of dual-phase steel. 2016
ISBN 978-3-7315-0513-6

Band 8 Vedran Glavas
Micromechanical Modeling and Simulation of Forming Processes. 2017
ISBN 978-3-7315-0602-7

Band 9 Eric Bayerschen
Single-crystal gradient plasticity with an accumulated plastic slip: Theory and applications. 2017
ISBN 978-3-7315-0606-5

Band 10 Bartholomäus Brylka
Charakterisierung und Modellierung der Steifigkeit von langfaserverstärktem Polypropylen. 2017
ISBN 978-3-7315-0680-5

Band 11 Rudolf Neumann
Two-Scale Thermomechanical Simulation of Hot Stamping. 2017
ISBN 978-3-7315-0714-7

Band 12 Mauricio Lobos Fernández
Homogenization and materials design of mechanical properties of textured materials based on zeroth-, first- and second-order bounds of linear behavior. 2018
ISBN 978-3-7315-0770-3

Band 13 Malte Schemmann
Biaxial Characterization and Mean-field Based Damage Modeling of Sheet Molding Compound Composites. 2018
ISBN 978-3-7315-0818-2

Band 14 Jürgen Albiez
Finite element simulation of dislocation based plasticity and diffusion in multiphase materials at high temperature. 2019
ISBN 978-3-7315-0918-9

Band 15 Maria Loredana Kehrer
Thermomechanical Mean-Field Modeling and Experimental Characterization of Long Fiber-Reinforced Sheet Molding Compound Composites. 2019
ISBN 978-3-7315-0924-0